Science Education in the Early Roman Empire

Richard Carrier

Pitchstone Publishing
Durham, North Carolina

Pitchstone Publishing
Durham, North Carolina
www.pitchstonepublishing.com

10 9 8 7 6 5 4 3 2 1

Library of Congress Cataloging-in-Publication Data

Names: Carrier, Richard, 1969-
Title: Science education in the early Roman Empire / Richard Carrier.
Description: Durham, North Carolina : Pitchstone Publishing, [2016] |
Includes bibliographical references and index.
Identifiers: LCCN 2016027638 | ISBN 9781634310901 (pbk. : alk. paper)
Subjects: LCSH: Science, Ancient. | Science—Europe—History. |
Science—Study and teaching—Europe—History. | Rome—Civilization. |
Rome—Social conditions.
Classification: LCC Q124.95 .C367 2016 | DDC 507.1/037—dc23
LC record available at https://lccn.loc.gov/2016027638

Cover design by Alex Gabriel

CONTENTS

1. Introduction

In preparation for my forthcoming book *The Scientist in the Early Roman Empire* it became necessary to fill a gap in the literature on ancient education. Past treatments of that subject have glossed over the question of science education specifically, at most providing a few cursory paragraphs of largely superficial analysis, sometimes even confessing it to be inadequate. Accordingly I had to provide a thorough analysis of the evidence on my own for my Columbia University dissertation, and the resulting chapter there is here expanded as an independent monograph.[1] The present book provides a generalized survey of the entire ancient education system as it was in the early Roman Empire, wherein the significance of science in its curriculum is emphasized.[2] And by "science" in this context I mean knowledge of the natural world, actively pursued by empirical means, with an ongoing concern for developing and employing valid methods of drawing conclusions from observations.

1. Richard Carrier, "Attitudes Toward the Natural Philosopher in the Early Roman Empire (100 B.C. to 313 A.D.)," dissertation (Ph.D.), Columbia University (Department of History), completed in 2008. The remainder of that dissertation is currently being expanded into my full-length treatment of the Roman scientist.

2. Note that in the present volume I will frequently cite (in addition to leading scholarship) standard academic reference works like the *Oxford Classical Dictionary* (*OCD*) and the *Dictionary of Scientific Biography* (*DSB*) (see bibliography for a complete abbreviation key), not only for the benefit of lay readers, but also because they contain summaries and bibliographies of the state of the field of interest even to experts, and many entries even professional readers might not know are in them, and it's worth reminding them of these valuable resources, especially for topics outside their usual field of study.

The thesis I shall explore in my next book on the subject of ancient science is whether ancient scientists (then known as "natural philosophers") were marginalized or perceived with indifference by the educated or empowered elite. Many historians have claimed they were, and then argue that this attitude explains the failure of the ancients to realize the great Scientific Revolution hailed in modern times. Both claims are dubious. Even more dubious is the claim that the seminal change responsible was the introduction of Christian values in the education system, paving the way for the ensuing revolution.[3] Nevertheless, in pursuit of all three claims, an important indication of social attitudes toward the natural philosopher is that science and natural philosophy were not considered very important in early Roman education.[4] Other achievements were regarded as far more valuable, particularly the ability to speak well in public and in court, and to write with an impressive literary style. As we shall see, these were almost the exclusive aims of all Roman education except at the highest levels of instruction, and even then only a select few pursued studies that involved a broader acquaintance with science and natural philosophy. This does not entail any actual hostility toward natural philosophers and their work, but it does imply a certain level of indifference toward them, since literary and verbal achievements were more valued.

These conclusions can be misleading, however. Any comparison with modern education, for example, is not entirely apt. Science has gradually become a required and more widely pursued subject of modern education at all levels, but in most respects this is a *product* of the Scientific Revolution and its subsequently observed benefits, which obviously had not transpired in the ancient world. In a sense the elevation of science in modern education reflects the same values that motivated the ancients: then as now, the education most widely embraced in a society emphasizes the learning that is most likely to produce a good living or maintain an elevated social status, since it is by mastering certain subjects that one can enter a good career or avoid derision as an unlearned 'hick'. In antiquity the subjects that could

---

3. Most competently argued in Hannam 2009; but for a refutation of key elements of his case (and, more directly, of all less competent defenses of the same thesis) see Carrier 2010. See also Efron 2009 and Gruner 1975.

4. In this book I shall use "Roman" to refer to all Greek or Latin speaking inhabitants of the Roman Empire (whether actual Roman citizens or not).

gain these ends were literature and eloquence, not science or mathematics, or even a merely practical literacy, unlike today. This does mean the ancient social system rewarded classical literacy and eloquence far more than science or mathematics, almost the exact reverse of the situation now, but this does not entail that science and mathematics had negative or no value, any more than erudition or eloquence have negative or no value today.

Any attempt to attribute the 'lack' of a scientific revolution in antiquity to an apparent absence of value for science in the prevalent educational system must find a more relevant comparison. And for that, historians of the Scientific Revolution need to know how widely scientific subjects were actually taught in primary and secondary education before the 17th century. Studies of medieval university curricula do not apply to that question, since few among the population of the time would ever have advanced so far, if they received any education at all.[5] So if medieval primary and secondary education did not significantly differ in subject material from their ancient counterparts, or in the numbers thus educated, the question then becomes whether and to what extent late medieval higher education differed from ancient higher education in these respects. And it is not likely to come out well by comparison. The actual science content of medieval universities was very limited and circumscribed, consisting almost entirely of a limited survey of Aristotle (already scientifically obsolete even in the early Roman era) and (for medical students) Hippocrates and Galen, and the methods taught were predominantly not empirical.[6] In any school there was typically only one mathematics professor (who was not even an

5. On the nature and origins of the university system, which arose after the 11th century, see Ferruolo 1998, Pedersen 1997, De Ridder-Symons 1992, Hastings 1987, Bowen 1975, and Haskins 1923 (with one possible exception in the East, the rather unique Academy of Constantinople, founded in the 5th century, developed in the 9th, and disbanded in the 14th: see Markopoulos 2008, Constantelos 1998 and, though perhaps less reliably, Kyriakis 1971). For a broad, although fairly unsophisticated introductory survey of the development of education from Greece, through Rome, into the Middle Ages and then the modern era, see Dobson 1932. For a broader multicultural survey of ancient education, see Bowen 1972.

6. See Rossi 2001: 192–202. It should also be noted that medieval students did not enjoy the same intellectual liberty that ancient students did (largely due to attitudes detailed here in chapter seven). See Freeman 2002, esp. where supported by Carrier 2010: 419 n. 56.

engineer and rarely even a productive astronomer) for every dozen or so professors of medicine (none of whom taught any hands-on dissection, apothecary, or experimental methods). Improvements in these respects only began in the 16th century, and thus again were more a product than a cause of a then-ongoing Scientific Revolution. In fact, universities did not conduct or promote new research, and to the end of the 17th century no actual scientific discovery was ever made in a university. What we take for granted now in educational aims and standards is in fact largely a product of the mid-20th century, when the nuclear and space programs launched science and engineering as pervasive nationalist concerns. And though in America a slow and limited development toward increasing science content in primary and secondary schools had begun in the late 18th century, that's again well *after* the Scientific Revolution—and even then it was only in the later 19th century that any science content could be called typical in American schools.[7] Indeed, even in England, often credited with driving the Scientific Revolution as we know it, as late as the dawn of WWII "the British government and educational systems treated applied mathematics and statistics as largely irrelevant to practical problems. Well-to-do boys in English boarding schools learned Greek and Latin but not science and engineering, which were associated with low-class trades,"[8] which description would just as well fit schools in the United States before the American Revolution. It's unlikely we'll find it to have been much different in the Middle Ages. Or after. A rising craze for mathematics in the 16th and 17th centuries is often noted, for example,[9] but standing back and looking at European society as a whole, this appears only to have occurred in rarified circles, just as in antiquity. The only difference is that far more sources have survived from the later period, creating only an illusion of more interest and activity.

---

7. Reese 2005.

8. McGrayne 2011: 63. Evidence of this largely dismissive attitude toward science and mathematics in modern educational systems prior to the atomic bomb can be found throughout the 18th, 19th, and early 20th centuries. Evidently the aristocratic attitude underlying it was not diminished even by the Scientific and Industrial Revolutions: e.g. D. Lee 1973: 70–71 and Green 1990: 470–73 and 855 (notes 38 and 39).

9. e.g. Nahin 1998: 8–47.

Though a more focused analysis of the actual facts of medieval education must be left to another study, the other half of this comparison shall be provided here: what was the Roman educational system really like in these regards? We will then compare pagan and Christian educational ideals, to see if any stark differences remain, because any subsequent change in Christian values (such as during the Middle Ages) cannot be attributed to Christianity in any original sense. In fact such values may represent only a *recovery* of what had already been pagan values before Christianity monopolized Western culture.

2. Who Was Educated

There was no formal educational system in the ancient world, but there was an 'educational system' in the broader sociological sense. Within the Roman empire, a more or less standard set of curricula advancing through several levels of educational achievement was almost universally available to those who had the time and money for it. But as Hènri Marrou aptly put it, "in any society a high degree of culture is the privilege of the favored few, and in all ancient societies, which were highly aristocratic in form, these favored few were very few indeed."[10] Obviously, then, the question of how much natural philosophy entered general education during the Roman period is superseded by the question of how much of the population received any education at all. Since most people received none, it follows that most people received little or no exposure to science and natural philosophy. Although Greco-Roman culture provided a notable exception to this assumption in the popularity of public declamation, which I'll discuss in a later chapter, the scientific content of this entertainment was not pervasive. Education mattered a great deal more. But even within the smaller subset of the population who acquired at least some education, most of *them* would not advance very far. Just like today, at every stage higher on the ladder of education one can expect to find fewer and fewer people.[11] With all this in

---

10. Marrou 1956: 427 (= Marrou 1964: 593 n. 12, "dans toute société, la haute culture n'est le privilège que d'une élite; et dans toutes les sociétés antiques, si aristocratiques, cette élite à toujours été peu nombreuse").

11. See Cribiore 2001: 1–12, who starts with the analogy of climbing a difficult hill in Lucian's *Hermotimus*: the higher you go, the fewer who continue.

mind we need to ask two questions in this chapter: who received an education in the Roman period and how much science and natural philosophy would that education have conveyed to them? The latter question will receive a more detailed answer as we proceed through the following chapters.

First, the Roman population was divided into slave and free, and in most cases slaves would only receive the education their owners paid for, which admittedly did happen a lot, since literate and educated slaves were more valuable and always useful.[12] But given that in all agricultural societies the demand for manual labor always far exceeded the demand for secretaries and other educated professionals, it is reasonable to assume that only a small proportion of slaves were educated, though possibly greater than the proportion of the free who were, especially among the landworking poor, since an educated slave was worth more, and there may have been more among the wealthy willing to finance the education of a slave than of a free person. Second, among the free there was a fading distinction between actual Roman citizens and noncitizen inhabitants of the Roman empire. That distinction was outright eliminated by Emperor Caracalla when he made citizenship universal in 212 A.D.[13] But even before that there is no clear sense in which Roman citizens had any greater access to education than non-citizens. There were a variety of municipal educational charities (which I'll discuss in chapter eight), where having the *local* citizenship of a particular metropolis could improve a freeman's access to education, but that remained the case even if they weren't Roman citizens specifically. Roman citizenship alone had only the advantages of lower taxes—with one possible exception: the thousands of native Italians who benefitted from a special welfare program instituted by Emperor Trajan at the dawn of the

---

A similar analogy is used in the pseudonymous philosophical dialogue *Tablet of Cebes*, translated and discussed in Seddon 2005, where the path toward truth and happiness is a fabulous mountain of education and insight, yet the higher you go, the more who give up the climb (see also Trapp 1997 and Diogenes Laertius, *Lives and Opinions of Eminent Philosophers* 2.125). The *Tablet of Cebes* is commonly dated to the 1st century A.D. (though it purports to have been written centuries earlier), although the Christian Tertullian claims a close relative of his composed it (Tertullian, *Prescription against Heretics* 39), which if true (and if he means the same book) would more likely place it in the 2nd century A.D.

12. Harris 1989: 247–48, 255–59.

13. Sherwin-White 1973: 279–87, 392–93.

second century A.D., and the beneficiaries of similar charities established in Italy and beyond by private benefactors and subsequent emperors.[14] These did not provide for education specifically, but might have made education more accessible by supplementing the income of poor families, especially in Italy. But in the end, one's status as a slave or citizen did not matter with regard to educational opportunity as much as gender or wealth—whether one's own wealth or that of one's family, patron, or owner.

It's worthwhile to analyze the ancient educational system from the perspective of women, as with a good understanding of their situation, the far more privileged situation of men comes into sharper relief. Though women were not formally or legally excluded from education at any level, resources and opportunities for education were distributed in favor of men.[15] Given the prevailing prejudices, assumptions, and socio-economic realities of the time, families would sooner scrape pennies to send a son to school than a daughter, and this preference would become more acute the higher the educational level concerned. Conversely, the wealthier the family the less pressing such choices would be. But in any case, women were expected to concentrate on raising families, while men were more welcome in education

14. See Ramsay 1936, Duncan-Jones 1982: 288–319 and 333–42, Patterson 1987: 124–33, and Woolf 1990 (and sources in each). For a summary see *OCD* 61–62 (s.v. "alimenta"). See also Marrou 1964: 437 with 611 n. 11 (= Marrou 1956: 303) and Lewis & Reinhold 1990: 2.255–59, 2.268–70 (§II.70). For similar charities outside Italy (which may have included non-citizens): C.P. Jones 1989. See also Lamotte 2007 for the Trajanic system (possibly conceived by Nerva: Page 2009) and Cao 2010 for the whole gamut of alimentary charities in the Roman period, public and private. Pliny the Younger, *Panegyric* 26–28, reports that in Rome alone such a scheme was serving "nearly five thousand" boys (and the number was growing: Duncan-Jones 1982: 290, 293), and this may have been in addition to girls. Even if girls were not already included, they certainly were by the time of Antoninus Pius in the middle of the second century, as reported in the *Historia Augusta* = 'Julius Capitolinus', *Life of Antoninus Pius* 8.1, and confirmed by coins and reliefs, cf. Cohen & Rutter 2007: 66–67, but considerable evidence suggests this practice had already begun with Trajan, cf. Ramsay 1936. It is estimated that between one hundred thousand and two hundred thousand Italian children benefitted at any given time from Trajan's charity (Duncan-Jones 1982: 317), while even more would have benefitted from other similar charities in and out of Italy.

15. But on the rising access to education among women in the Hellenistic and Roman periods see Pomeroy 1977: 52–53 and 1995: 170–76; and Whitmarsh 2001: 109–16.

at higher levels, or else had a far greater opportunity to materially benefit from it. The latter point is particularly noteworthy. As of the 1st century A.D., women could not advocate in court and were apparently not much appreciated as public orators or philosophers, hence they could not benefit from a rhetorical or philosophical education in the way a man could.[16] For instance, when Quintilian has occasion to list examples of rhetorically educated women, all of them precede him by more than a century. The most recent was Hortensia, who delivered an oration during the triumvirate (43 – 32 B.C.) that was so impressive, Quintilian says, that it was still read in schools.[17] There is also an inscription at Delphi that appears to honor a woman for her oratory in the 2nd century A.D.[18] But these were certainly exceptional. We do not hear any examples of women making a career out of public speaking.

Female philosophers are likewise rarely heard of, though not unknown, being far more common than female orators.[19] Though only writing at the end of the 3rd century, when the Roman social system was already in decline, the Christian educator Lactantius reports the fact that Epicureans, Platonists, and Stoics had long wished slaves and women to receive an education in philosophy, but had never succeeded in realizing

16. See Bauman 1992. In the early first century A.D. Valerius Maximus compiled a list of exceptional cases of women advocating at trial (*Memorable Deeds and Sayings* 8.3, cf. also 3.8.6 and 8.2.3) and it was possibly in response to such behavior that around the same time the standardized 'praetorian edict' began to include a prohibition against women representing others in court (though still allowing them to testify as witnesses and to represent themselves in court). Related prejudice is indicated by Gaius, *Institutes* 1.190–91, who reports a prevailing view that women are naturally incompetent. Although Gaius dismisses the sentiment as unfounded (as we'll later see others did), it was evidently still a common view.

17. Quintilian, *Education in Oratory* 1.1.6. For his date and background see *OCD* 1251–52 (s.v. "Quintilian (Marcus Fabius Quintilianus)"). On the women he names as orators see Snyder 1989: 123–27 and Plant 2004: 104–05.

18. Agusta-Boularot 2004: 322, 330, with *Fouilles de Delphes* 3.4.79.

19. Levick 2002; J. Barnes 2002: 293–94 and 303; Taylor 2003: 173–226; Irby-Massie 1993; Snyder 1989: 99–121; Waithe 1987; Pomeroy 1977: 57–62; Marrou 1964: 578 n. 39 (= Marrou 1956: 414–15); Tod 1957: 140; *OCD* 1577 (s.v. "women in philosophy"); and following notes.

this dream.[20] This cannot be really be true. Lactantius falsely claims, for example, that only one woman was ever educated in philosophy in the whole of human history—and yet he names Themistoclea but not, for example, Hipparchia or Arete.[21] At the same time he claims only one slave was ever taught philosophy in the whole of human history, which is certainly untrue—again, he names Phaedo but not, for example, Epictetus. Both remarks are therefore hopeless exaggerations. But even exaggerated, this at least entails that Lactantius personally did not know of many female philosophers—at least, they could not have been common or else his hyperbole would have

20. Lactantius, *Divine Institutes* 3.25.

21. Diogenes Laertius, *Lives and Opinions of Eminent Philosophers* 6.96–98 (Hipparchia) and 2.86 (Arete). On Hipparchia there is also an epigram by Antipater of Sidon (*Palatine Anthology* 7.413). Pythagoras was also reputed to have made philosophers of his wife and daughter (Pomeroy 2013; *OCD* 988 and 1450, s.v. "Myia" and "Theano"; however cf. *EANS* 49 and 781–82, s.v. "Aisara of Lucania" and "Theano, pseudo") as well as Themistoclea, and Plato's "disciples" included "two women," Lastheneia and Axiothea, according to Diogenes Laertius, *Lives and Opinions of Eminent Philosophers* 3.46; and others (Snyder 1989: 106–13; Plant 2004: 68–86). Plato himself claims one woman (Diotima) among the 'presocratic' philosophers (insofar as she was teaching when Socrates was a student, assuming the account is not fictional): Plato, *Symposium* 201d–212b. Epicurus was also famed for including women in his school—elite prostitutes in particular (these *hetairai*, lit. "companions" or "lady friends," were already expected to be well educated in order to hold stimulating conversation with elite male clients, cf. Pomeroy 1995: 89–92, 141, and thus would have been of all women at that time the most suited to studying philosophy). The most famous of these was Epicurus' lover Leontion (herself perhaps the first published feminist philosopher—her tract defending women against the disparaging remarks of Theophrastus is lost), but other women studying under Epicurus had telltale names suggestive of a similar profession (Hedeia, "Sweety"; Mamarion, "Titsy"; Erotion, "Sexy"; Boidion, "Oxeyes," similar to our "Doe Eyes"; and the more ordinarily named Demetria, "Demeter's Girl"); see Pomeroy 1995: 103–05. A Galenic treatise also praises an otherwise-unknown Arria, identified as Galen's best friend (*philtatê*), and as a brilliant Platonist philosopher: cf. Nutton 2004: 223, citing *On Theriac to Piso* Kühn 14.218 (not 14.208 as misreported in Nutton), and Nutton 1997 (which convincingly defends Galen's authorship of this treatise). Notably an unnamed female Platonist philosopher is also the dedicatee of Diogenes Laertius' *Lives and Opinions of Eminent Philosophers* 3.47, written around the same time. For many more examples see scholarship cited in previous and following notes (e.g. women philosophers as dedicatees of inscriptions: Levick 2002: 134, etc.).

appeared ridiculous.[22] He also fallaciously conflates being a philosopher with merely being taught philosophy, so his remark obscures whether he knew of any female *students* of philosophy. There must have been many women who studied some philosophy without going on to teach it or write about it (as we'll certainly see in later chapters).

Though female orators and philosophers were rare, other occupations that required an education specifically in scientific subjects, such as medicine and engineering, seem not to have accepted many women in practice, even though again there was no formal prohibition. It is notably during the late Hellenistic and early Roman period, however, that we start to see evidence of female doctors—not merely educated midwives, which had long been known, but actual surgeons and physicians.[23] The same evidence confirms this was not a common sight, but their existence entails that some women could receive considerable education in the sciences, even if only in exceptional cases. So it is not surprising to find Galen (in the late 2nd century A.D.) endorsing Plato's view that women are as intellectually capable as men in all the sciences as long as they have the same education, while even the

22. Or perhaps not: Lactantius evidently had no awareness of how ridiculous his claims could be, for example denying well-established science of his time proving the earth is a sphere by scoffing at the notion of there being upside down people on the other side of it: Lactantius, *Divine Institutes* 3.24. Though this was a notion shared by the uneducated (who were the vast majority of the population), his position and argument would both have been laughable to most educated persons of his time: on both points see Pliny the Elder, *Natural History* 2.65.

23. Kudlien 1970: 17–18; Nickel 1979; King 1986; Jackson 1988: 86–87 and 1993: 85–86; Irby-Massie 1993: 364–67; Nutton 1995: 18–19; Künzl 1995 (see also occasional data in Gourevitch 1970, supplemented by Agusta-Boularot 2004: 328–29, esp. n. 61); Parker 1997; Nutton 2004: 142, 196–98; and most recently Flemming 2007 (see also *EANS* 94, s.v. "Antiokhis of Tlos"). Flemming also examines the question of whether any women wrote medical books, but finds the evidence disputable, encountering the same problem that plagues the alchemical tradition (see note below). Of course, whether we know of any is not the same as whether there were any. In any case see *EANS* 121, 173, 281, 316, 354, 447, 456, 482, 500, 552, 564, 588, 596, 719, 725, 755, 778–79 (s.v. "Aquila Secundilla," "Aspasia," "Elephantine/Elephantis," "Eugeneia" and "Eugerasia," "Hagnodike of Athens," "Iuliana," "Iunia/Iounias," "Kleopatra of Alexandria," "Laïs," "Metrodora," "Muia, pseudo," "Olumpias of Thebes," "Origeneia," "Romula," "Salpe (of Lesbos?)" and "Samithra/Tanitros (?)," "Soteira," and "Thaïs").

ideal midwife described and recommended by Soranus appears hardly less skilled and educated than many scientific doctors of his time.[24] Nevertheless, such enlightened attitudes toward women in antiquity resembled more those of the 19th century than the 20th.[25] There is otherwise only one known example of a female astronomer (Hypatia) and only one known female expert in harmonic science (Ptolemaïs), and one other mathematics professor (Pandrosion) who would have had some understanding of mathematical sciences (since in antiquity one did not study mathematics apart from science, as such a distinction only arose much later).[26] There is

24. Galen, *On the Doctrines of Hippocrates and Plato* 9.3; Soranus, *Gynecology* 1.3–4.

25. Cf. Nutton 2004: 235.

26. Ptolemaïs of Cyrene wrote a treatise on harmonics and music theory around the turn of the era (first century B.C. or A.D.). We know nothing else about her, except that her work appears to have been at least modestly brilliant and influential (Levin 2009: 230–93; Plant 2004: 87–89; Irby-Massie & Keyser 2002: 344–45; Barker 1989: 239–42; *OCD* 1234 [s.v. "Ptolemaïs of Cyrene"], *NDSB* 5.172–73 [s.v. "Ptolemais of Cyrene"], and *EANS* 705–06 [s.v. "Ptolemaïs of Kurene"]). Hypatia of Alexandria, a professor of Platonic philosophy in the late fourth and early fifth century A.D., wrote commentaries in mathematics and astronomy, and was consulted on the construction and use of laboratory instruments for the study of physics (Deakin 2007; Dzielska 1995; Snyder 1989: 113–20; *DSB* 6.615–16 [s.v. "Hypatia"]; *NDSB* 3.435–37 [id.]; *OCD* 716 [id.]; *EANS* 423–24, [s.v. "Hupatia"]; and Harich-Schwarzbauer 2011). Pandrosion taught in the fourth century A.D. (Netz 2002: 197; *EANS* 608–09, s.v. "Pandrosion"), but we're told no details of her scientific interests. Likewise there may have been at least one female agricultural writer, but this conclusion is based on a single letter in a name that could have been corrupted in transmission (*EANS* 637, s.v. "Persis," which could be an error for Perses). Some alchemical treatises were attributed to otherwise unknown female authors, but their names do not seem authentic (e.g. "Maria," cf. *EANS* 531, more probably an apocryphal attribution to the sister of Moses), and fanciful pseudonyms were common in the alchemical tradition (Irby-Massie & Keyser 2002: 238–41, 243–45; Plant 2004: 130–47; e.g. *EANS* 446, s.v. "Isis, pseudo (Alch.)" and "Isis, pseudo (Pharm.)"), and since alchemists believed their art had been "revealed" to mortal women by fallen angels in their attempt to woo them (from the *Book of Enoch* 6–8; cf. *DSB* 14.631, in s.v. "Zosimus of Panopolis" and *OCD* 51–52, s.v. "alchemy"), suspicion is warranted when alchemical knowledge is attributed to a woman. But their involvement in the art is possible. Other possible female scientists in antiquity are listed in *EANS* 1029 (and discussed in their associated entries).

no evidence of female engineers, and there probably were none given the mechanical labor and inconvenient circumstances such a job would involve, and the associated social expectations, especially since engineers commonly worked in the army or on construction sites in positions of authority over men. Some women may have been able to avoid these barriers by carving out a narrower niche in mathematics (Pandrosion), astronomy (Hypatia), or harmonics (Ptolemaïs), since these required only working privately with instruments (and we know there were some female craftsmen in various trades). The fact that we know only one name from each science does not mean there were no others—to the contrary, odds dictate there must have been others, for us to have known of even these. But they were clearly rare. Similar problems and attitudes may have limited the number of women even in occupations that did not require as much training, such as secretary or librarian, or even schoolteacher, although real opportunities for women in some of these fields did exist.[27]

For these reasons and others, women did not have equal access to resources and opportunities for education in the Roman empire, though the disparity grew in proportion to educational level and shrank in proportion to wealth. In the first respect, social and economic realities would conspire to ensure that women would be increasingly excluded the higher they got in educational achievement, but there is ample evidence of many women in the lower stages of the Roman educational system, and of some in the higher stages. In the second respect, elite women were expected to have an education, especially by prospective husbands of status, and thus it would be rare for the daughter of an elite family to go without an education. This same fact would imply that among the middle classes a woman's prospect for social advancement through marriage could depend on having at least a modicum of this expected educational refinement. So there may have been a higher proportion of educated women among the middle class than of educated women among the poor, though still nowhere equal to the proportion among the upper classes.[28]

---

27. See Agusta-Boularot 2004, who finds evidence of female teachers above the elementary level scarce, but abundant for female scribes, secretaries, and librarians, and to some extent elementary teachers (ibid.: 329–30).

28. Critics sometimes protest at the supposed anachronism of applying this familiar three-class division to ancient society (e.g. Toner 2002). I shall prove it

A famous example is Julia Domna, wife of emperor Septimius Severus and mother of emperor Caracalla, who patronized and associated with philosophers and intellectuals for about twenty years (in the late 2nd and early 3rd century A.D.) and may have been a philosopher herself, though we hear no specifics of Julia's education.[29] Likewise queen Cleopatra, who in the 1st century B.C. was famed not only for speaking and writing numerous languages but also for hobnobbing with doctors and philosophers, evincing a considerable education.[30] Though these women represent the ultra-elite, the sources do not suggest their educational achievements were unusual for elite women generally. For example in the 2nd century A.D. the patroness of Nicomachus, described in the opening chapter of his *Manual of Harmonics*, was quite evidently well educated and keen to read books on harmonic theory. And many other examples can be found, both epigraphic and literary.[31]

An example of such an exceptional upper-class woman is Pompey's wife Cornelia, who was renowned for having studied literature, music, and geometry and attending lectures in philosophy.[32] This was clearly a point of admiration that made her more attractive. Plutarch further adds that

---

applicable in *The Scientist in the Early Roman Empire*, but it suffices here to say that if the 'upper class' consists of those who did not have to work to live (e.g. large landholders) and the 'lower class' of those who barely survive at or near subsistence level (e.g. day laborers and menial slaves), there remains a clearly identifiable population in between (e.g. those who must work to live but who generate such a surplus that they can send their children to school or erect expensive epitaphs), who have so much in common with any familiar 'middle class' (in both means and values) as to obviously warrant the label (see Atkins & Osborne 2006: 4–11).

29. She is called "the philosopher Julia" in Philostratus, *Lives of the Sophists* 2.622 and was said to have actively studied philosophy in Cassius Dio, *Roman History* 76.15.7 and 78.18.3; both men knew her personally. For scholarship on Julia Domna see Bowersock 1969: 101–09, Hemelrijk 1999: 122–28, Levick 2007: 107–23, and OCD 754 (s.v. "Iulia Domna")

30. Plutarch, *Antony* 27–29.

31. Evidence of education among elite women is comprehensively surveyed in Hemelrijk 1999, with Levick 2002 discussing women's educational access to philosophy in particular; several prominent examples from the Roman period are discussed in Snyder 1989: 122–51 (and more in Levick 2002: 146–48 and Plant 2004).

32. Plutarch, *Life of Pompey* 55.1–2 (cf. Plant 2004: 101–03).

Cornelia did not have the "annoying inquisitiveness" that other women do who undertake such studies, which entails Plutarch was familiar with many similarly educated women in his own day. The fact that he found their excessive queries unladylike could simply reflect a prevalent male chauvinism in Roman culture, but it could also reflect the fact that women did not get as much opportunity to study intellectual subjects as men, and hence on the rare occasions they were permitted to, they tried to get the most out of it—resulting in behavior Plutarch uncharitably perceived as annoying. Juvenal likewise reveals there were many educated women in high society who were fully bilingual in Latin and Greek and knew literature, philosophy, rhetoric, and mathematics, and it's worth pointing out that he lambasts such women only when they showed off or argued too much or otherwise behaved unseemly for a woman, so he does not otherwise attack the notion of women being educated.[33] So the remarks of both Plutarch and Juvenal confirm it was not rare but nevertheless uncommon for women to be educated to a fairly high level, although even when they were, men still expected women to behave as they should, an attitude that would have limited both encouragement and opportunities for educational advancement.

Nevertheless, Plutarch expected cultured women to study geometry and astronomy and at least a little philosophy:

> For a woman studying geometry will be ashamed to be a chorusgirl, and she will not accept the stench of magic charms if she delights in the words of Plato and Xenophon. And if anyone claims the ability to call down the moon, she will laugh at the lack of sense and education of the women who believe these things, not being unschooled in astronomy herself.[34]

He adds that an educated woman will have heard the story of the witch Algaonike who tricked women by claiming to be able to darken the moon when she already knew an eclipse was coming.[35] Hence Plutarch's point is that educated women will be forearmed against such cons. Though this cannot have been usual, Plutarch clearly did not imagine his advice to be unrealistic. He was serious in his insistence that women must not be

---

33. Juvenal, *Satires* 6.186–88, 6.434–56, and 6.574–78.
34. Plutarch, *Marriage Advice* 48 (= *Moralia* 145b-d).
35. For more on this Algaonike see Bicknell 1983.

allowed to have their minds infected with nonsense, and the best vaccine against this was a good education, ideally as good as her husband's—which tells us something about the education *men* were assumed to have.[36] He also argues that a woman will think better of herself and behave better, too, if she partakes of "what the Muses bring and bestow upon those who admire education and philosophy."[37] That a man could educate his wife in natural and moral philosophy, or court a woman already so educated, must have reflected reality on some scale, or else Plutarch's advice in this regard would have sounded ridiculous to the ears of anyone who read or heard it (and unlike Lactantius, Plutarch was not as accustomed to saying ridiculous things). It might not have been routine among the elite or the aspiring upper middle class, but it could not have been bizarre either. Still, this did not mean many women actually attended rhetorical schools or spent much time attending philosophy lectures—the education Plutarch imagines required only a modicum of exposure to the arts (as I'll show in subsequent chapters), which could have been completed before a woman reached marriageable age. Nevertheless, Plutarch was not alone in perhaps hoping for more. In the first century A.D. the revered Stoic sage Musonius Rufus specifically argued that women were as competent as men and ought to receive the same education, up to and including philosophy, which for any Stoic would have included natural philosophy.[38]

Though women among the wealthy and reasonably well-to-do could find nearly equal educational opportunities as men, at least up to a certain level of study, those less well-off would be in a completely different situation. As already noted, insofar as a family could afford any education at all for their children, they would be far more likely to invest what they had in educating their sons rather than their daughters, for all the above reasons.[39] But few among the poor would be able to send even a son to school, and this would include most freed and freeborn families in the Roman empire. A pseudonymous treatise on education written sometime during the early

36. Plutarch, *Marriage Advice* 48 (= *Moralia* 145e).

37. Plutarch, *Marriage Advice* 48 (= *Moralia* 145e-146a).

38. Musonius Rufus, *Sermons* 3 and 4. On the educational ideals of the Stoics see chapter seven.

39. See Harris 1989: 239–40. A reality illustrated even in the case of orphans: Cribiore 2009.

Roman empire recognizes that there was a concern for the education of the poor, but it does not offer a solution. "My dearest wish," the author says "would be that my scheme of education should be generally useful," but "if some, being needy in their private circumstances, shall be unable to avail themselves of my directions, let them lay the blame therefore upon fortune and not upon him who gives this counsel." In other words, when the question "What about the poor?" did come up, the effective response was, "Oh, they're just screwed. That's not my problem." As the author says, "Even the poor must endeavor, as well as they can, to provide the best education for their children, but, if that be impossible, then they must avail themselves of that which is within their means."[40] Thus, it was simply expected that families had to pay for their children's education out of their own pocket, or manage somehow on their own. As we shall see later, there were some exceptions in the form of state, community, and private foundations for the education of children among the more successful Hellenistic city-states, many of which survived into Roman times, while similar charities were on the rise (or being restored) under the early Roman empire. But the extent of their reach has been questioned. Such foundations were certainly not universally available. In the communities that had them, they would have expanded the pool of children who had access to education, but they did not necessarily guarantee that every child got in. There are indications that parents still had to find the means to pay some of the expenses, and those expenses would multiply with the number of children, and increase again at every stage of education.[41]

After surveying all the available evidence, including the presence and nature of these educational charities, William Harris concludes that access to education was still greatly limited even to the free, much less the slave population. Since literacy is essential for "really large quantities of precise information" to be passed on, especially "to minds that were by no means exceptional" (which would describe most even among the elite), any significant dissemination of scientific lore would have been effectively impossible without first disseminating a basic literacy to the populace.[42] But "the classical world, even at its most advanced, was so lacking in the

---

40. Pseudo-Plutarch, *On the Training of Children* 11 (= *Moralia* 8e-f).

41. Harris 1989: 130–34, 242–47, 283, 307.

42. Harris 1989: 82.

characteristics which produce extensive literacy that we must suppose that the majority of people were always illiterate" and therefore, for the same reason, we can expect the majority did not learn much about science or natural philosophy.[43] This does not mean they absorbed nothing of these subjects, since oral lore would disseminate some knowledge (although through many distorting filters). But real theoretical understanding of any scientific fact or field would be unlikely. As Harris notes, modern-scale literacy "could not have happened" in the Roman period "without radical changes in the educational and social systems and some unimagined change in the technology used for reproducing the written word."[44] His point being that such a scale of literacy as is seen today is not possible without a massive state or social infrastructure that was clearly not present in antiquity, and is very unlikely without certain innovations in technology that were also not available at the time, like the printing press, the erasable pencil, cheap paper, or even the chalkboard (though wax tablets, potsherds and other kinds of trash, and sandboxes, are widely attested as serving the same function in antiquity).

The well-to-do among the Greeks—those of "gymnasium" status or the social equivalent and above—did value and strive for universal literacy within their class and probably nearly achieved it.[45] For them it was a matter not just of necessity, but of status and prestige. This was especially the case for those in positions of significant power or influence, where an uncultured man could become a target of ridicule or social exclusion or be passed over for favor or advancement. The situation was somewhat similar among the Latin colonies and territories—even if not as pervasively, men of the equivalent "curial class" and above were probably literate, the more so the nearer to Rome, in either geography or civic prosperity.[46] But below the level of the "gymnasial" or "curial" classes, however—in other words, for the considerable majority of the population—the situation was very different. Among artisans, who comprised the social equivalent of a 'blue collar'

43. Harris 1989: 13.

44. Harris 1989: 282.

45. Harris 1989: 276–77. On the "gymnasial" class as a recognized elite social status see Whitehorne 1982 and Hin 2007. Where students were taught is discussed in the next chapter.

46. Harris 1989: 264, judging especially from evidence at Pompeii.

middle class, many, Harris says, "took a negligent or wholly uninterested attitude towards the education of their sons" and consequently were only occasionally literate themselves, and probably only as much as was practical or useful.[47] And again, many slaves were literate, but in proportion to the total slave population, only very few, those lucky enough to be groomed for managerial, secretarial and administrative positions, in and out of government. And far fewer of those would have had the time, means, and inclination to pursue advanced studies. Like the majority among the artisan class, most slaves surely settled for, or only received, what met their practical needs.[48] Finally, outside cities and sizable towns, literacy was rare and limited even for the higher ranking residents of any village or country district, due to the diminishing availability of teachers and resources.[49]

Overall, on more than reasonable grounds Harris estimates that even in the most literate cities, regions, or provinces, fewer than 15% of the population during the early Roman Empire was reasonably literate, and fewer than 5 or 10% among the remainder.[50] Even for Italy, "the evident feebleness of the school system and the general shortage of interest in financing it" as well as "the lack of imperatives which might have led the well-to-do to take an interest in the education of the free-born poor," plus "the evidence for the illiteracy or semi-literacy of some well-to-do freedmen," all indicate that literacy was by no means an achievement of the majority, and these conditions would only have been worse outside areas long under the sway of Greco-Roman cultural influence.[51] In the Latin West "there was no special effort to educate more than a very few of the local population" and though Greek populations of the East more firmly embraced an ideal of widespread literacy, there were only isolated attempts

47. Harris 1989: 277 and 264. Similar indifference to education may have been common in many occupations. For example, in *Advice for an Epileptic Boy* 3–4 (= Kühn 11.361–62, see Temkin 1934), Galen takes it for granted that all gym teachers (*paidotribês*) are uneducated (*apaideutôn*), even though he assumes studying under them was a regular part of every boy's education.

48. See Haines-Eitzen 2000: 55.

49. Harris 1989: 278.

50. Harris 1989: 266–67, 272.

51. Harris 1989: 266.

to produce it.[52] Ultimately, while the elite could be expected to have achieved universal literacy among themselves, and "in every city some of the more ambitious merchants and artisans also came to be literate or semi-literate," the majority of city populations would not fall into either category, while in the provincial countryside, "literacy is unlikely to have extended much beyond landowners and their specialized slaves."[53] Of course it was unlikely to be any different in the Middle Ages.

Despite attempts to argue otherwise, the data simply do not support more optimistic numbers.[54] For example, a recent team of scholars has argued from surviving tax records that the ratio of grammarians to children in Egypt averaged 1:165 in the 3rd century B.C. and conclude that this confirms "widespread diffusion of Greek education in the Egyptian countryside, against Harris' more pessimistic view."[55] But that does not follow. As they note themselves, many of those counted as teachers were not actually teachers but teachers' wives and dependents, and most of the actual teachers in those records would have been teaching in cities and major administrative centers, not "in the country."[56] And even for the general populace 1:165 is not that impressive—by comparison, the ratio of teachers to children in the U.S. is in the vicinity of 1:15.[57] Any ancient ratio also cannot be overgeneralized. The economic realities of antiquity must be

---

52. Harris 1989: 272–82.

53. Harris 1989: 269.

54. To this point I have depended on the undeniably thorough analysis of William Harris. The only significant attempt to respond to Harris has been Humphrey 1991, which attempts to offer rebuttals or qualifications to Harris' conclusions, but the included essays do not offer any effective challenges to his methodology and I found no evidence there that Harris hadn't already considered, either in fact or in kind. See also in support of Harris *OCD* 843–44 (s.v. "literacy"), Hezser 2001, Woolf 2000, and Johnson and Parker 2009 (which also includes an extensive post-Harris bibliography: 333–82).

55. Clarysse and Thompson 2006: 2.125–33 (with Katelijn Vandorpe). Cited in Oleson 2008: 735.

56. Clarysse and Thompson 2006: 2.125, 129. See also Thompson 2007: 129–31. And this is before the Roman era, when possibly a fifth of the populace of Roman Egypt lived in cities (Tacoma 2008).

57. According to the Center for Education Reform: http://www.edreform.com/Fast_Facts/K12_Facts.

taken into account. For example a country estate may have one teacher in its employ for the children of its wealthy landowners, but that does not mean the twenty villagers next door were enjoying a teacher-child ratio of 1:20. For example, Clarysse et al. found a teacher-child ratio of 1:10 for (only) the Greek inhabitants of the town of Lagis, which had a total population of around 600, but Greeks there were a small minority, and they may have been recently settled families of considerable wealth. In fact, there were only 21 adult Greek men in Lagis, only 2 of whom were registered as teachers, and possibly one of those was the dependent of the other.[58] If there were only one wealthy family in Lagis, their distance from a major urban center would necessitate hiring a teacher for their children, who may have been the sole teacher in Lagis, and who may have taught the children of a couple of other families (as many as could afford it), but this in no way means all the Greek children there were being educated, much less the whole population. The resulting 1:10 ratio may therefore reflect nothing more than what was typically enjoyed by wealthy elite families, which is wholly unsurprising. When we look at the numbers more broadly, there were over 300 children in Lagis, and possibly only 1 actual teacher, which would make a dismal ratio of no better than 1:300. Even if all the 20 or so *Greek* boys of Lagis were being educated by that one teacher, this would more likely be because they were the children of rich landowners or successful professionals, and therefore already exceptional. And even then, at our most optimistic, the literacy rate in Lagis would have been only 20/300, or 6⅔%. Which quite confirms Harris' estimates. Literacy simply wasn't widespread. But it remains correct to say that literacy did grow more widespread under the Roman Empire, and "a far greater proportion of the population of the Roman empire could make use of texts than was the case in most ancient societies."[59]

There is also no doubt that the ancient elite greatly valued education. In the 1st century B.C. the Latin engineer Vitruvius cites the wisdom of Theophrastus, and the example of a shipwrecked Aristippus, and other authorities, to argue that an education is more useful than money, because it cannot be lost, seized, looted, or stolen—but more importantly, an educated man is "a citizen in every country," for he will be welcome everywhere and will always find employment or other like-minded gentlemen who will

58. Clarysse and Thompson 2006: 2.131.

59. Johnson and Parker 2009: 46–51.

take him in.[60] Among the Greeks, writing near the same time as Vitruvius, Diodorus recorded a legendary law of the 6th century statesman Charondas ensuring that all children of his community would be taught to read and write. Diodorus cites the example with elaborate praise, although (and even if mythical) the context and tone suggests it was praiseworthy partly because it was exceptional—it represented an ideal that might have sometimes been met, but not an actual universal practice.[61]

Hence when we find observations of the actual state of things among the populace of the time, we see most remained without even basic literacy, much less any more significant education. The 3rd century Christian scholar Origen, for example, observes that most people are in fact uneducated, arguing "it is necessarily the case that among the multitude of those overcome by the gospel, many times more are commoners and more 'hick' than those who are trained in reason," because, Origen says, that's how the population breaks down, not because Christians focus on the 'hicks'.[62] Likewise, both Origen and Lactantius note that most people work such long hours just to get by that they do not even have time for an education.[63] Tertullian says "the simple people, not to mention the ignorant and inexpert, are always the greater part of believers"—*always*, because they are the greater part of the population being evangelized.[64] Galen likewise observed in the late 2nd century that the "average man on the street," who would often be "goatherds, cowherds, diggers, harvesters" and so on, were

60. Vitruvius, *On Architecture* 6.pr.1–3.

61. Diodorus Siculus, *Historical Library* 12.12.4–12.13.3.

62. Origen, *Against Celsus* 1.27: *hoi idiôtai kai agroikoteroi*, "idiots and farmhands," or more literally, "nonprofessionals and countrydewellers," the latter in the comparative ("more so" hence "more hick"). In context these words carry the definite connotation of "ignorant laymen and those more rustic," compared with *tôn en logois gegumnasmenôn*, "those practiced in reason," i.e. those having received oratorical education and experience in the public *gymnasia*. See *LSG* 15 (s.v. "agroikos"), 819 (s.v. "idiôtês" III.1–3), and 362 (s.v. "gymnazô" I.Pass.) and 1057–59 (s.v. "logos" e.g. IV.1).

63. Origen, *Against Celsus* 1.9–13; Lactantius, *Divine Institutes* 3.25.

64. Tertullian, *Against Praxeas* 3: using the words *simplices*, *imprudentes*, and *idiotae*, "simple, naive," "foolish, ignorant," "layman, amateur," respectively. See *OLD* 1764–65 (s.v. "simplex" 8.b), 853 (s.v. "imprudens" 1), 820 (s.v. "idiota" 1).

illiterate and uneducated.[65] There were no doubt many urban occupations Galen could have named that were equally illiterate, like the cooks, cobblers, carpenters, dyers, wool-workers, wool-carders, weavers, and bronzeworkers whom he elsewhere implies were typically uneducated.[66] So, too, Pliny the Elder assumes "country folk" are illiterate, and Ptolemy expects the same of farmers and shepherds.[67]

The 1st century pagan educationalist Quintilian attacked the belief that the masses are "so slow of understanding that education is a waste of time and labor" on them.[68] And Cicero argued, against a popular contrary opinion, that even "virtual farmboys" can learn astronomy.[69] But although both Cicero and Quintilian said this prejudice against commoners is groundless, and that the uneducable are actually rare, the fact that the contrary belief was common enough that they had to dissent from it confirms that most people were, in fact, not educated at all, nor expected to be.[70] Which means most people in the Roman Empire had no formal exposure to science education, either. But as we shall see in coming chapters, apart from the most isolated rural folk, the illiterate were nevertheless likely to know *something* about

65. Galen, *On the Affections and Errors of the Soul* 2.3 (= Kühn 5.71): *tois epitugchanousin anthrôpois* literally translates "the men chanced upon," which in context indicates the average man you would meet if you just grabbed someone at random. Notably, all of Galen's examples (*aipolois*, "goatherds"; *boukolois*, "cowherds"; *skapaneusi*, "diggers"; and *theristais*, "reapers, harvesters") are agricultural, but these would still have been the most common occupations in antiquity, even among men who would be wandering around town during the day. Galen says such men are *agumnastoi*, lacking an education of the *gymnasia*, but in context he clearly means lacking any education at all.

66. Galen, *On the Therapeutic Method* 1.1.5 and 1.3.2.

67. Pliny the Elder, *Natural History* 25.6.16 (where *agrestes* are *litterarum ignari*); Ptolemy, *Tetrabiblos* 1.2.7–8 (*geôrgos* and *nomeus*).

68. Quintilian, *Education in Oratory* 1.1.

69. Cicero, *On the Republic* 1.15.23–24, using the example of Gaius Sulpicius Gallus (which I shall discuss in *The Scientist in the Early Roman Empire*) who claimed he had taught the legions under his command the astronomical cause of a lunar eclipse. Emperor Claudius attempted something similar on a wider scale (see chapter eight).

70. Hence several times Quintilian refers to the assumed illiteracy of the lower classes (e.g. *Education in Oratory* 2.20.6, 2.21.16, 10.3.16, 12.10.53).

science, just not very much, and very little that was correct even by the standards of the time. Meanwhile a significant minority of the population did get some degree of education. And yet, again, on any of these measures the Middle Ages are unlikely to have been significantly better, but more likely worse.

After her meticulous study of the evidence Raffaella Cribiore came to the same conclusion, that few people made it even beyond elementary education, much less completed any course of higher education, and even from the start "the pool of starting students was incommensurably smaller" than today.[71] Whether indeed the Middle Ages saw any difference in this respect is for another study to determine. But for antiquity, even by the most optimistic estimates, we have most of the population receiving no education at all—at least 80%. Of the 20% or fewer who received any education, many would not advance beyond rudimentary literacy and numeracy, and it is reasonable to expect that no more than half would go on to pursue studies at the secondary level (beyond, perhaps, vocationally, which was an educational dead end). That would leave at most 10% of the population. Of those who got that far, many would continue no further—again, it is reasonable to expect no more than half did, so less than 5% of the population would continue to higher education. Of them, no more than half again would likely undertake a proper philosophical education, since the other opportunities available at that point would be sufficient to satisfy most ambitious students, so most would only go on to pursue law or advance no further. That means those who would move ahead to the equivalent of graduate studies in any scientific or science-related field would be fewer than 2% of the population—probably much less (considering that science professors would typically require students to have completed the more advanced curriculum at the secondary level which we will examine in chapter five). Therefore, while even in the most literate cities no more than 1 in 5 people of the time could have enjoyed the practical benefits of a basic literacy and numeracy, it is likely that fewer than 1 in 10 advanced to any level of education where some exposure to science or natural philosophy would even be possible, and probably fewer than 1 in 50 would ever expose themselves to any significant amount of science-related educational content. And at that level we would find mostly free men, very few women or slaves,

71. Cribiore 2001: 1–3.

although the numbers of the latter may have been greater among those who took the more science-rich curriculum at the secondary level (a possibility we'll examine in chapter five).

Yet these are excessively optimistic numbers. More realistic estimates would be closer to 10% (1 in 10) finishing only primary education, 2% (1 in 50) finishing only secondary education, 0.4% (1 in 250) undertaking any kind of higher education, and 0.08% (1 in 1250) undertaking studies with significant science content (whether at the secondary or advanced levels). Even the latter number would consist mainly of philosophy students whose interest in science would be divided and often exceeded by an interest in ethics or logic. Beyond that, as we shall see, education in an actual *science* (like medicine or engineering or even just a serious passion for natural philosophy in general) required special preparatory studies at the secondary level that most students at that stage did not pursue, so the actual pool of potential *scientists* would be far smaller still—and then, of course, the pool of *actual* scientists would be even smaller, and most of them would become mere practitioners or professors rather than conducting original research (which is the case even today, as most doctors and engineers do not advance their fields). If this 5:1 progression was maintained (which, though reasonable, is only speculation), that would mean 1 in 6,250 Romans had the opportunity of undertaking a scientific occupation; 1 in 31,250 actually did; and 1 in 156,250 went on to conduct original research that advanced their field.

If, as several scholars estimate, the average population of the Roman empire was 60 million, which means on average 40 million over the age of 15, roughly 20 million of whom were men, then by these highly speculative figures there would have been about 128 research scientists throughout the empire at any given time. That seems excessively high given extant records, which in turn implies the progression curve was even steeper than 5:1, which entails that access to scientific knowledge in ancient education may have been even more exclusive than just speculated. For example, Netz employs a different method to estimate the number of productive investigators in the *mathematical* sciences alone and arrives at a result of much less than 100 at a time.[72] However, these final estimates should not be mistaken for the

72. Netz 2002: 201–09. See also Rihll 2002: 12–21 who discusses how various aspects of the education system limited the number of scientists in all eras of

number of *practicing* scientists, who, like today, know the science and use it but don't contribute to its advance. For instance, the number of scientific doctors must have been well into the hundreds at any given time merely to account for their ubiquitous presence in extant records, and likewise there must always have been hundreds of engineers to serve all the legions, attend to all the building demands of Roman ports and cities, and maintain the massive hydrological infrastructure of the empire, and the 5:1 progression curve would predict this total number of practicing doctors and engineers as being over a thousand, which fits expectation. But even if we imagine one or two thousand such scientists in every generation, that still equals a small proportion even of the elite.

---

antiquity, though she comes to no definite conclusion as to numbers. Nutton 2004: 153 estimates (and on good grounds) very large numbers of practicing 'doctors', but most of these would not have been well-educated or engaged in research (since any "healer" might have made it into his count). At the other extreme, Collins 1998: 76–77 estimates the number of 'significant philosophers' (those responsible for major innovations) at no more than thirty in any given century, but his methods rely on extant literature, which can only have resulted in an undercount.

3. What They Were Taught

Having established "who" was exposed to ancient education, we need to examine "what" they were exposed to, and how much it involved science or natural philosophy. Though there were no formal names for different stages of education as there are today, and they overlapped a great deal, it's not unreasonable to apply familiar terms by analogy.[73] The ancient system can be broken down into primary, secondary, higher, and advanced education—although these divisions were not clearly made at the time, we can see something like them in practice. Ancient primary or 'elementary' education usually began at or soon after the age of seven, continued into the pre-teens, and involved only the imparting of literacy and simple numeracy (basic counting and arithmetic). Ancient secondary education (perhaps closer to what Americans would call middle school) typically occupied the early teens and always involved more advanced grammar and basic literary studies, and—for some who chose to undertake a more complete course of education and had the time and money for it—geometrical and other

73. The following analysis draws on the findings and conclusions developed in Cribiore 2001 and corroborated in the scholarship that will be cited in more specific detail as the occasion arises. For a brief yet broad survey of ancient education see *OCD* 487–91 (s.v. "education, Greek" and "education, Roman") and König 2009. Marrou 1964, once the standard resource, has been updated considerably: see Too 2001 and Pailler & Payen 2004 (which also includes a bibliography of books on ancient education published after 1964 on pp. 361–68), as well as Wolff 2015 and Sandnes 2009: 16–39; and Bloomer 2011 (for imperial education in Latin). A handy if eclectic collection of sources on ancient education is also provided in Joyal, McDougall, and Yardley 2009.

studies under a separate teacher (which I'll discuss in chapter five). Many who received a primary education did not continue—not least because the fees for secondary teachers could be four times higher than a primary teacher.[74] And of those few who completed secondary school, many of *them* would not continue, thus narrowing even further the number who went on to higher education. Those who wanted to enter a lower-level occupation such as scribe or accountant could receive specialized training at the secondary level—the ancient equivalent of vocational school—and would need to advance no further. But of those who did go on to higher education, there were options.

Ancient higher education frequently began in the mid to late teens and usually consisted of an education in declamation and debate, in other words, "oratory" or "rhetoric." It was possible to skip rhetoric and begin a philosophical education instead, so in a sense philosophy was another form of higher education (a choice loosely analogous to the modern university distinction between studying the humanities or the sciences), but it was common to complete some degree of rhetorical education before switching to philosophy, and many philosophy professors included training in

---

74. Diocletian's *Edict on Maximum Prices (EMP)* 7 (some of which is in Lewis & Reinhold 1990: 2.425–26; with relevant discussion in Harris 1989: 308) established maximum (not typical) prices in a time of extreme inflation at the end of the 3rd century, but these still indicate the relative social value of different teachers. Prices are in (inflated) denarii: elementary teacher (*paedagogus* and *magister litterarum*), 50d per student per month (*EMP* 7.67–68); elementary math teacher (*calculator*), 75d (*EMP* 7.69); shorthand teacher (*notarius*), 75d (*EMP* 7.70); scholar or librarian (*librarius* and *antiquarius*), 50d (*EMP* 7.71); Greek or Latin grammarian or geometer (*grammaticus* and *geometres*; the latter would include all teachers of advanced mathematics), 200d (*EMP* 7.72); professor of rhetoric or philosophy (*orator* or *sophistes*), 250d (*EMP* 7.73); and teacher of engineering (*magister architectus*), 100d (*EMP* 7.76; Clarke rightly notes that this does not mean an engineering teacher held a lower status than a grammarian, it simply reflects the fact that an architect would not be a mere lecturer, but already earning a living, and even employing his pupils as apprentices in carrying out his other paid work, cf. Clarke 1971: 113–14). Notably, there is no *magister medicus* or *magister astronomus* listed, although these might have been assumed under *sophistes* and *geometres*, respectively. For some idea of the actual values, the maximum wage for a farm laborer was only 25d per day (*EMP* 7.1), which would amount to a monthly take of between 500d and 750d, about what an elementary teacher would earn from having ten to fifteen students.

rhetoric in their curriculum.[75] Either way, a study of philosophy had to be chosen out of a student's own interest (or his parents' or patron's insistence), and the reality is that most students who continued into higher education completed or undertook only rhetoric. After that, advanced education marks the ancient equivalent of a "graduate degree," which in antiquity took the form of a professional education in medicine, engineering, or some other academic specialty—law being the field most commonly chosen (even more so than today).

Since there was no state regulation of the professions, one could pursue training as a doctor, lawyer, engineer, or whatever, by skipping steps in the educational process (e.g. rhetoric and philosophy), but this would not likely produce a prestigious career. The real money to be made came from the elite, and in the competitive market of the time, a cultured professional had an enormous advantage. All the evidence suggests that a Roman gentleman was unlikely to hire someone whose lack of polish made him come off as an incompetent hick, especially when he could hire someone who had that admired polish. Hence doctors, engineers or lawyers who could demonstrate their extensive eloquence and erudition, and exhibit a sharp, practiced mind, would obviously do well against their weaker competitors. Thus, a mix of rhetorical and philosophical education was essential for those who intended to do well at an advanced occupation. For the same reason, the best teachers in the field would not likely take students who did not come to them adequately prepared. In both respects Galen's treatises *On the Therapeutic Method*, *On Examinations by Which the Best Physicians Are Recognized*, and *Exhortation to Study the Arts* all reveal this standard, and even attempt to improve upon it (although I am certain rhetorically exaggerating how often members of the elite ignored it).

In the previous chapter I concluded the number of practicing scientists in antiquity would have been, at most, in the low thousands in any given generation, which was a small proportion even of the elite, much less of the general populace. Those scientists will certainly have received as good a science education as there was to be had at the time (and I'll explore their education in chapter seven). But for every one else who managed

75. For a general introductory discussion of ancient "higher" education, including rhetoric, philosophy, and the *enkyklios paideia*, and the ages of students embarking on it, see Kleijwegt 1991: 116–23. I'll discuss these studies in more detail in coming chapters.

an education, as we shall see, the focus was almost entirely on developing first a basic, then a cultured literacy, and little else. In Cribiore's words the "idols of the educated public" were orators, not scientists or natural philosophers. Moreover, ancient education was built on "subservience to conventional values" that included "no attempt to question the transmitted doctrine."[76] Teresa Morgan likewise concludes that "what they taught, at any given level, recurs again and again in the surviving evidence in remarkably similar forms across vast geographic distances, a wide social spectrum and a timespan of nearly a thousand years" and yet, "skilled professions such as medicine are absent from texts to do with literate and numerate education, and the natural assumption is that they were regarded as a separate area of education," one not widely pursued.[77] Though there were many ardent proponents of a broad education including contact with the sciences, and such an education was available to those with sufficient interest and means, such an ideal curriculum simply is not what most students undertook.

Another issue affecting exposure to educational content was language. Even throughout the early Roman empire the language of science remained Greek by both tradition and practical necessity, for at least two reasons. First, all past scientific work was in Greek—many centuries worth—and thus to build on the past, even to critique it, required the ability to read it, and to interpret and comment on it with considerable competence. Second, as we know from the repeated complaints of Latin authors like Lucretius and Cicero, Latin lacked a sufficiently precise and developed vocabulary in the sciences, yet neologism was considered distasteful among many of the Latin litterati.[78] But the Greeks had already developed ancient sciences to an advanced level, and in the process had produced precise and specialized vocabularies in every field. Thus it was simply easier to learn Greek than to completely overhaul the Latin language just to play catch-up—easier

76. Cribiore 2001: 239, 247.

77. Morgan 1998: 3, 6.

78. Pliny the Younger, *Letters* 4.18; Pliny the Elder, *Natural History* 2.13.63; Lucretius, *On the Nature of Things* 1.136–39, 1.830–33, 8.258–60; Quintilian, *Education in Oratory* 8.3.33; Cicero, *On the Boundaries of Good and Evil* 3.51 and *Tusculan Disputations* 2.35. For qualifications and discussion of this point see Fögen 2000, Brunschwig 2002, Dufallo 2005, and (most importantly) Ostler 2007: 118–219.

especially since the Latin elite had already fully embraced strong motives to master Greek for several other reasons (from the political to the aesthetic), while even those outside the elite had frequently to deal with Greeks in the south of Italy, on the coast of France, in Sicily—and in trade, everywhere. Therefore, the procedure of bilingualism would obviously have been more useful and efficient than a massive program of revising the Latin language and producing mass quantities of accurate translations into Latin of centuries of accumulated scientific works in Greek.[79]

Insofar as the wealth and traditions of the elite remained strong, this presented no great difficulty. A bilingual education for the children of a wealthy Latin family was simply a matter of pride and course.[80] And Greeks were already masters of the relevant language.[81] But the lower down the social ladder someone fell in the Latin West, the more a need for bilingual education would grow into a barrier to good scientific knowledge. Elizabeth Rawson correctly notes that though in the early Roman period the elite held to a tradition of strong bilingualism, even then "it is likely that it was harder, and more expensive, to get a good education and buy books (in two languages) than it was in the Greek world, and this put the rich, as well as the members of their households and their protégés, at an advantage."[82] Indeed, to learn a sufficient degree of both Latin and Greek required two

79. In later periods, in the West, as Greek education declined, translation was almost the only option left for rescuing scientific texts, and as a result very few were saved—most recovered texts in the Renaissance had to be imported. The gradual replacement of *Latin* during the Renaissance with the languages of new rising empires (e.g. English, Italian, French, German) was another case of lingual economy, of a rather different kind. Although inhabitants of the Roman Empire had long been not merely bilingual but multilingual: Mullen & James 2012.

80. See, for example, the casual observations of Quintilian, *Education in Oratory* 1.1.12–14. Adams 2003, and Adams et al. 2002 provide detailed discussion (superseding Horsfall 1979, whose evidence is mostly pre-empire and whose analysis ignores comparative studies of modern bilingualism). For further context and bibliography: *OCD* 231–32 (s.v. "bilingualism"). And on the Roman adoption of Greek-style education in general: Wallace-Hadrill 1983: 26–49.

81. Although ordinary Greeks would still have to learn the technical vocabulary of the sciences, which would present an obstacle of its own, though not as great (it's easier to learn new words in an already familiar language, and technical dictionaries were available, e.g. Horsfall 1979: 81–82; Witty 1974).

82. Rawson 1985: 98.

separate teachers, which doubled the cost to parents, while the expected preparatory studies (for advanced scientific education or simply for becoming a well-rounded student) required even more teachers and thus even greater expense. Quintilian, speaking from twenty years of personal experience, clearly thought it unlikely that a student could master all the required preparatory subjects (like geometry, music, and astronomy, as well as Greek and Latin) unless he learned them more or less simultaneously, and when he was young and had no other duties.[83] Yet to procure all these teachers at the same time would have magnified the expense to a family many times over.

On the one hand, this meant that any advanced education in science was only realistically possible for those who mastered Greek, which meant the pool of available personnel for scientific study would be far smaller among Latin populations. At the same time, this meant those who were neither native speakers of Greek nor wealthy or driven enough to procure a fully bilingual education, if they received any exposure to scientific content in lower education at all, or even through lore and hearsay, it would be through the filter of frequently inaccurate, imprecise, or incomplete translations into Latin, which would limit both the quantity and quality of the knowledge passed down to them. Hénri Marrou rightly identifies this fact as one of the causes of the rapid decline of science in the Christian West relative to the Christian East after the 3rd century A.D.[84] We can see the effect of this decline in science education by comparing the relative quality and sobriety of detail in Pliny the Elder's *Natural History* or Seneca's *Natural Questions* with the excessively superstitious and often wildly inaccurate content of their 7th century equivalent, Isidore of Seville's *Etymologies*—even though all three works were in Latin by Latin authors who were all laymen in the sciences yet otherwise among the most learned men of their respective times.[85] In contrast, many of the Greek scientific encyclopedias of the East

83. Quintilian, *Education in Oratory* 1.12.6.

84. Marrou 1964: 372–88, with 592–94 notes 11–17 (= Marrou 1956: 254–64, 426–27); also argued in Greene 1994: 30 and documented (though with excessive rancor directed at the pre-Christian period) in Stahl 1962 and 1971, and more soberly in Diederich 1999.

85. Clagett 1955: 146–67. This does not mean Pliny and Seneca were immune to the same faults, especially as they were laymen, but the difference is a matter

retained more sober and accurate detail, preserving in somewhat better form the achievements of the past, although still imperfectly, incompletely, and unimaginatively.

Hence one can rightly say that under the Christians science stagnated in the East, but actually declined in the West. The responsible circumstance was both cultural and economic. After the collapse and decline of the Western economies, fewer and fewer people could afford the time or expense to finance a bilingual education (much less a massive translation campaign of advanced works requiring mass neologism), while at the same time more and more among the middle or lower classes, and even from the ranks of barbarian tribes, rose to positions of power and influence without a full or bilingual education, breaking the tradition of higher social expectations. At the same time, since the split of the empire between East and West fell largely along linguistic lines, and a Latin translation of the Bible was already fully embraced, both the need for and interest in mastering Greek continually declined. As we'll see in chapter nine, a certain degree of anti-elitism among the Christians might have contributed to the effect, who often saw pursuit of the trappings of Hellenism to be somewhat vain and pretentious (a point I'll further demonstrate in *The Scientist in the Early Roman Empire*), and ultimately unnecessary when a perfectly serviceable language for evangelizing the flock was available. Though some of the Western elite continued to maintain a personal or antiquarian interest in Greek, their numbers dwindled, and ultimately there ceased to be any strong political, economic, or aesthetic motive for the medieval upper class to master Greek. In fact, insofar as bilingualism was thought at all urgent, so-called 'barbarian' languages had become politically and economically more important. The same motives would have become even weaker still among the middle and lower classes of the Latin West. All this placed a

---

of degree, and that difference was enormous (particularly given that the much-maligned Pliny the Elder actually had his facts right a lot more often than has been assumed: e.g. French & Greenaway 1986; Healy 1999). Isidore wrote around the turn of the 6th and 7th centuries, but Boethius already exemplifies the same decline earlier in the 6th century: *DSB* 2.228–36 (s.v. "Boethius, Anicius Manlius Severinus"), *OCD* 238 (s.v. "Boethius, Anicius Manlius Severinus"), and *EANS* 195 (s.v. "Anicius Manlius Seuerinus Boëthius"). For similar examples of consequent decline in the same period see Beagon 1992: 52–53, and Stückelberger 1988: 111–26, 179–84.

dark and distorting window between the medieval populations of the Latin West and good scientific knowledge, since whatever they had of it *only* came to them through the filter of often inaccurate, imprecise, or incomplete translations—which could only have been compounded by a corresponding decline in the availability and quality of education generally, and of scientific values in particular.[86]

But in our period of interest, none of this had happened yet, and bilingualism was still a strong tradition and the common standard among the Latin elite. But even at Rome's height, the disparity that would consume the West in the Middle Ages would already have been felt across the class divide, at least to some degree. The Latin-speaking poor and working class would not likely have been sufficiently bilingual to enjoy good access to scientific knowledge in any era, and thus would have been at a disadvantage in this respect, compared to poor and working class Greeks. The significance of this disparity might not have been great, since plenty of nonsense and error trickled down among Greek populations, and ignorance was still the rule, but it shall become clear below that access to good scientific knowledge, through actual education or the casual attending of public lectures and orations, would have been easier and readier at hand for those already facile with Greek than for those who could not afford to be. But even then dissemination of scientific knowledge remained limited.

This conclusion cannot be carried too far, however. For all the reasons surveyed in this chapter, access to science content in the ancient education system was certainly limited by a number of social, economic, and cultural factors. Yet the exact same disparities and limitations are likely to have obtained even during the Renaissance (and were likely even greater in the Middle Ages than they were in antiquity), so whether this even amounted to a difference between the two periods is questionable. But we must put that question aside for now and move on to discuss how much of science the average Roman student would encounter in school.

---

86. On the role in all this of a declining Latin-Greek bilingualism see Ostler 2007: 58–104, 203–04, 211–12, 246–49.

4. Lower Education

Very little science content would be found at the lower rungs of ancient education.[87] But the basic tools needed to access such content were provided. Early in the 4th century the Christian scholar Eusebius observed of pagan society, "as for knowledge of things visible to the senses, they did not think it necessary to disseminate this very much among the multitude, nor to teach the majority of the populace the causes of the nature of what exists," except, perhaps, as much as might have been needed to promote a basic gist of creationism.[88] Though he made this observation somewhat late for us to be sure of its pertinence before the changing and chaotic circumstances of the 3rd century A.D., it probably reflects the universal reality in all periods of Roman history. As far as we know, the Greco-Roman elite made no effort to "educate the masses" in scientific fact or theory (what Eusebius means by the 'causes' and 'nature' of what exists, and by a knowledge of what's 'plain to see'). Though that may be obvious of Roman *society*, the same accusation could be leveled against the Roman educational system itself: those who entered it were already the privileged or fortunate minority, and yet even they were not presented with much in the way of scientific fact or theory. Insofar as any philosophy trickled down into standard lower education, it was predominantly moral philosophy, logic perhaps second to that, but natural philosophy least of all. As an example, note the general absence of natural philosophy as an interpretive or educational concern in Plutarch's treatise

87. On ancient primary and secondary education in general see Kleijwegt 1991: 75–91 and Cribiore 2001.

88. Eusebius, *Preparation for the Gospel* 11.7.10.

How the Young Man Should Study Poetry, which reveals a commonplace emphasis on moral over natural knowledge among pagan educators.[89] Likewise, the pseudonymous treatise *On the Training of Children* says children should get a taste of every branch of learning, but no more than a taste, and though philosophy should be given more attention, this primarily meant, above all else, moral philosophy, with a secondary emphasis on political philosophy. Natural philosophy is thus slighted again.[90]

The first stage of ancient education, the equivalent of elementary school, was entirely occupied with basic elements of literacy and numeracy: just reading, writing, and simple arithmetic.[91] That ancient primary education included arithmetic has been discussed in more detail by Serafina Cuomo, who concludes that "people who were literate were as a rule also numerate—that is, able to perform the basic arithmetical operations," although students who were planning to enter a mathematical trade (and such trades existed at all social levels, from accounting to philosophy) would study more mathematics under a specialist.[92] But apart from such basic numeracy, literacy remained the all consuming interest. This emphasis left no room for any other subjects to enter the curriculum. The second stage of ancient education, the equivalent of modern middle school or the early years of high school, expanded on this literacy with more in-depth study of literature, mostly in the form of poetry or verse, all with the aim of imparting a basic cultural toolbox that would prepare the student for rhetorical school or

89. Plutarch, *How the Young Man Should Study Poetry* (= *Moralia* 14e-37b).

90. Ps.-Plutarch, *On the Training of Children* 10 (= *Moralia* 7c-8a; moral philosophy: 7d-f; political philosophy: 8a)

91. Cuomo 2012. Cribiore 2001 documents the obsessive focus on reading and writing (on numeracy: 180–83). On the Roman *calculator* (arithmetic teacher) see Clarke 1971: 46–47 and Marrou 1964: 599–600 n. 13 (= 1956: 431).

92. Cuomo 2000: 46–47. On mathematics in Roman education generally see: Marrou 1964: 265–79 (= 1956: 176–85) and Rawson 1985: 156–69. For the broader context of the place of mathematics in the early Roman empire see Cuomo 2001: 143–211; and for the evident widespread need of basic numeracy and practical and applied mathematics in civic life (for which its inclusion in general education must have been essential) see Karin Tybjerg's survey in Oleson 2008: 777–84. Even just the process of paying one's taxes required it: see e.g. Wallace 1938 and Nelson 1983. As well as the ubiquitous employment of coinage, weights and measures: Oleson 2008: 759–77.

at least provide the ability to hold one's own in cultured company. This training included thorough familiarization with a more or less fixed set of classics, combined with some basic skills of interpretation and composition. This was the most common experience. The more ambitious students of wealthier parents would at the same time embark upon additional studies at this level, in preparation for higher education in rhetoric or philosophy. But because these special studies entailed additional time and expense and were seen as what we today would think of as 'college prep' courses, only a small subset of students at this stage pursued them. We will discuss this aspect of ancient education in the next chapter. Our present concern is with the content of education that would be experienced by all students at the secondary level.

Society did not hold elementary teachers in high esteem, and though a widespread value for the education they provided kept them pervasively employed, Harris has argued that the lack of dedicated buildings "symbolized the lack of interest in elementary education on the part of both society in general and the authorities in particular."[93] Such a conclusion, however, is a bit hyperbolic, since the availability of buildings was never really a problem. Dedicated school buildings may have been rare or even nonexistent, but though Harris doubts it, evidence does suggest students were taught in the lecture halls, campuses, and porticoes of public gymnasia, as well as in museums, libraries, temples, civic porticoes, and even public forums and basilicas, which were used at other times for public business—and, of course, in teachers' houses (on a model similar to the effective use of homes as churches and catechetical 'schools' by Christian congregations well into the 4th century).[94] In fact the ancient practice of multi-purposing its buildings is arguably more efficient. Nevertheless, elementary teachers were not highly esteemed and received little in the way of direct public support.[95]

93. Harris 1989: 237.

94. On the role of physical gymnasia as both schools and social institutions in the Hellenistic and Roman world see Brenk 2007, König 2005: 45–72, Kah & Scholz 2004 (esp. 103–28), Gauthier 1995, Delorme 1960: 316–36, and Forbes 1945. On the subject of school buildings in general, see summary in König 2009: 392–95 and König 2005: 45–49; on the Christian use of private homes for churching and teaching see MacMullen 2009: 1–10.

95. Harris 1989: 236–38. See Robinson 1921 for a still-useful survey of literary evidence for the social and economic status of Roman schoolteachers; and Laes

In contrast with elementary teachers, the status of teachers at the secondary level—the grammarian or schoolmaster, but also the geometry teacher—was substantially greater, and the status of teachers of higher education—primarily the rhetor, sophist, and philosopher—could be high enough to warrant actual prestige. Elementary teachers did not impart scientific lore to their students (although some trivial amount might have been conveyed simply in passing). But secondary teachers did have occasion to impart such knowledge in a small degree. Though this was particularly true of the geometer, we shall discuss that later, since only a minority of secondary students troubled to study that subject. All other students at this level would only learn of science through incidental and piecemeal commentary on the poets. Yet anyone who had received a good science education would be less likely to undertake an occupation as a grammarian, since if they had gotten that far they would likely also have, or easily be able to acquire, the skills to be a philosopher, if not an actual professional (whether, for example, a doctor, orator, engineer, astrologer, etc.). This would have the obvious effect of predominately placing teachers at the secondary level who had a less than adequate experience with science. So it is reasonable to assume that most teachers at this level would not know much if any real science, and whatever they did know, could easily have been wrong, inaccurate, incomplete, or mixed up with absurdities and nonsense.[96] Thus, the average student's exposure to science in secondary education would not only be limited, but could often be flawed and distorted as well.

Quintilian was aware of this problem and calls upon parents to ensure that their children find teachers who have a good grounding in at least basic science and other important fields. Since his advice suggests such qualified grammarians had to be sought out, they could not have been common, although Quintilian clearly believed a diligent parent could always find one, which entails teachers of such competence must have existed (at least in major cities).[97] He also says it was becoming increasingly common in his day for the introductory elements of a rhetorical education to be taught

---

2007 for epigraphic evidence (Kaster 1988 treats both but only for late antiquity). Most recently on their lives and social and economic status: Maurice 2013.

96. As observed by Morgan 1998: 3.

97. Quintilian, *Education in Oratory* 1.4–9.

by grammarians in the latter stages of secondary school.[98] This meant students at that level were getting practice in the more rudimentary skills of composition and argument, and Quintilian says this involved touching upon nearly every branch of knowledge.[99] Though he sees this as grammarians overstepping their bounds, since he thinks they are not up to the task, the practice clearly had become common enough to complain about. And although Quintilian says there may have been a few grammarians who were educated enough to manage it, this clearly meant most were not.[100]

Nevertheless, Quintilian expects a competent grammarian to be able to teach some rudimentary skills of speech and argument, and to be well versed in both poetry and prose.[101] In fact he insists that "every kind of writer must be carefully studied" by a grammarian, since he will need this in order to have a broad vocabulary and expansive knowledge for correctly interpreting the poets for his students.[102] Among other subjects, Quintilian specifically says a teacher cannot teach the poets if he is ignorant of astronomy, because the poets "frequently give their indications of time by reference to the rising and setting of stars" and make other references to astronomical knowledge and lore. Likewise, "ignorance of philosophy is an equal drawback" for the grammarian, since, among other things, "there are numerous passages in almost every poem based on the most intricate questions of natural philosophy." Even today we can see this was true of occasional passages in Homer or Virgil's *Aeneid*. Indeed, any avid student of Homer would tend to acquire a more expansive astronomical and anatomical vocabulary than even a typical modern American adult. And besides Homer, Quintilian specifically mentions Empedocles, Varro, and Lucretius as among those who "expounded their philosophies in verse," and this implies, first, that it could not have been too unusual for students to study Empedocles, Varro, or Lucretius, at least on occasion, and second, that even apart from the study of such rare poets, some oral instruction in science was an expected part of secondary education, even if not a very significant one. However, again,

98. Quintilian, *Education in Oratory* 2.1.

99. Quintilian, *Education in Oratory* 2.1.4.

100. Quintilian, *Education in Oratory* 2.1.6.

101. Quintilian, *Education in Oratory* 1.9.

102. Quintilian, *Education in Oratory* 1.4 and 1.9.

depending on the quality of teacher, the depth or quality of such scientific information might not have been commonly high.

According to Dionysius Thrax, the final goal of secondary education was "the critical study of literature," emphasizing a grasp of meter and literary devices, etymology, reasoning through analogies, and "notes on phraseology and subject matters."[103] The latter is what included at least minimal scientific content, though only incidentally. Quintilian also identifies the two primary tasks of the grammarian as *methodice*, imparting an ability to speak correctly and well, and *historice*, imparting an ability to interpret and understand what the student has read or heard, while adding a third focus, "instruction in certain rudiments of oratory," which meant imparting rudimentary skills in composition and argument.[104] Another educator, Asclepiades of Myrlea, likewise divided the responsibilities of the grammarian into three: *historikon*, *technikon*, and *idiaeteron*.[105] The first, *historikon*, involved "contextual and historical questions arising from literary texts," like mythology, geography, and glosses of special words, which is the context in which scientific knowledge might come up, though less commonly than other points of interest would.[106] In other words, at the secondary level, some science would have filtered into a student's education as notes and glosses on classical and popular poetry, but probably not much, and again perhaps not of much depth or quality.[107]

This was especially true given the fact that excerpts of Homer, Euripides, and Isocrates were the main "cultural package" at the primary level of education and continued to hold pride of place at the secondary level, yet those three authors are notably weak on anything like advanced scientific content, affording few opportunities for digressions on the state

---

103. Cribiore 2001: 185, quoting Dionysius Thrax, *Greek Grammar* 1.1 (c. 100 B.C.).

104. Quintilian, *Education in Oratory* 1.9.1.

105. Asclepiades of Myrlea (1st century B.C.) via Sextus Empiricus, *Against the Professors* 1.91–94 and 1.252–53.

106. Cribiore 2001: 186, 208–09. The "technical" (*technikon*) and "specialized" (*idiaeteron*) aspects of literary studies corresponded to the study of language itself (such as grammar) and the literary skills of exegesis, textual criticism, and aesthetic evaluation.

107. Cribiore 2001: 188–89, 205–10.

of contemporary science.[108] In fact, "no prose was read in a grammarian's class except for fables" and "didactic and moralistic" works by Isocrates, and various parables from Aesop and Babrius.[109] There was an occasional exposure to plays, but very little exposure to prose history (and apparently none to other prose works, such as scientific or technical literature), and therefore "almost no attention was paid to history and geography except for a wealth of minute information arising from specific points in the literary texts," and yet history and geography probably saw more treatment in this way than natural philosophy.[110] Nevertheless, in elucidating the poets of the time, many rudiments of astronomy, anatomy, botany, mineralogy and zoology must have been conveyed.

By studying recovered school texts (from papyri and ostraca) and analyzing the whole gamut of literary discussions of the educational process, Cribiore has been able to reconstruct, more or less, the typical grammar school curriculum. She found that it did expand the repertoire of authors studied beyond the canonical three (Homer, Euripides, Isocrates), most notably by adding Menander and Hesiod, but these additions were still weak on advanced scientific content.[111] Though Hesiod at least introduced more rudimentary discussion of nature (and especially astronomy) than any of the other authors, all such content was pre-scientific in respect to the advances in methods and discovery made since the development of natural philosophy as an actual field of research. And these details were the least popular part of Hesiod, who was studied more for poetic or religious value, and mythographic material, than anything else. At a more advanced stage of grammar school some attention was turned to Aeschylus and Sophocles, then Callimachus and Pindar, but these authors are just as weak on scientific content as the others. So, too, for other poets that students were more commonly exposed to at school.[112] Notably absent from recovered texts and curricula at both the primary and secondary levels of education are any works of philosophy, and not

108. Cribiore 2001: 179.

109. Cribiore 2001: 202.

110. Cribiore 2001: 247.

111. Cribiore 2001: 197–98.

112. Cribiore 2001: 198, 201–02; cf. 192–93, 202–04.

just natural philosophy, but even discussions of epistemology or logic.

It follows that Quintilian's expectation that grammarians needed to be able to expound on philosophical poetry did not reflect the common reality. Such a circumstance must not have been typical and might only have occurred in the classes of the most talented or rarest of school teachers. However, due to the existence of poetic treatises on astronomical subjects, which would have fit the predilection for studying verse, astronomy might have received more frequent attention in many grammar schools, particularly as it came up frequently even in the most studied of poems. It should also be noted that popular knowledge of rudimentary astronomy (knowing major constellations, the path of the sun, and similar basics) would have already been more commonplace (even among the illiterate) in antiquity than today, when people in general no longer look at the sky on any regular basis much less rely on its observation for navigation, timekeeping, and agriculture.

A papyrus from the pre-Roman period attests to grammar students having an astronomical work in which a student had jotted down notes, though the relative rarity of the find probably only confirms the general observation that such a deeper attention was not the typical experience in Roman schools.[113] Since ancient education was built on the expectation and practice of a strong exposure to only a few texts, mainly from a universally-embraced, quasi-canonical list, instead of a diverse exposure to many different texts, this would have greatly limited any rare exceptions to the standard course of study.[114] Most teachers had to cast a wide net to win students and build a livable income, which probably meant most did not want to alienate parents by introducing controversial philosophical claims and disputes at the secondary level. Instead, teachers, parents, and students preferred a fixed package of skills and texts that were universally embraced and thus comfortable ruts to roll in. Cribiore argues that this mindset, this preference for rigidity and subservience to widely shared expectations and values, dominated ancient education.[115] This would only have further marginalized natural philosophy as a subject of study in elementary and grammar schools, since it was often inextricably mixed up with various philosophical dogmas and controversies. Because disputed subjects were

113. Cribiore 2001: 189.

114. Cribiore 2001: 192–205.

115. Cribiore 2001: 204–05, 247–52; see also Morgan 1998: 240–70.

the last thing anyone wanted in schools—until students became suitably armed and mature, which was only expected to occur in the higher stages of education—in most respects natural philosophy, and thus science, had to wait.

There was one notable exception to this conclusion. The most popular astronomical poem was the Greek *Phenomena* of Aratus (early 3rd century B.C.), which also enjoyed many famous translations into Latin verse, including ones by Cicero (1st century B.C.) and Germanicus (1st century A.D.).[116] There is ample evidence that this poem found employment as a school text.[117] At least one school commentary on it has been recovered, explaining "cryptic astronomical lore," and there were others.[118] The *Phenomena* probably made its way into grammar schools not merely because it was in verse, but because it drew a lot from Homer, which made it more 'approachable' to underclassmen in school, who would already be familiar with Homeric content, style, and vocabulary. But like the astronomical content of Hesiod, the *Phenomena* contained relatively bland and obsolete science in comparison with the astronomical knowledge achieved up to Roman times, since it was only a versification of an old prose treatise by the early Platonic astronomer Eudoxus.[119] The *Phenomena* did not even discuss planetary theory (not even the obsolete model Eudoxus had developed), and yet that was perhaps the most scientifically important problem—and achievement—of ancient astronomy. Nor did the poem discuss research methodology or convey any particular sense of scientific values.

116. See Gain 1976 and Taub 2003: 51–54 and 2010; *DSB* 1.204–05 (s.v. "Aratus of Soli"); *OCD* 132 (s.v. "Aratus (1)"); *EANS* 123–24 (s.v. "Aratos of Soloi"). Numerous commentaries on the poem were produced (e.g. cf. Maass 1958).

117. Cribiore 2001: 142–43, 202. It is thus notable that the *Phenomena* of Aratus is the only work on science or natural philosophy ever quoted in the Bible, though only on a point of theology. See Clement of Alexandria, *Stromata* 1.19, discussing Acts 17:28, which depicts Paul speaking at Athens, although in fact this is the author (by later attribution, Luke) describing Paul speaking at Athens, so it could have been either or both men who were familiar with Aratus—or at least one verse, since the quotation Luke depicts Paul using could merely have been read off an inscription or epitaph or have entered common oral lore.

118. Cribiore 2001: 142–43, 202; Clarke 1971: 49–51; Bonner 1977: 78; Rawson 1985: 167; Marrou 1964: 273–74, 570 notes 11–12 (= Marrou 1956: 408).

119. Gee 2013.

And again, any instruction a grammarian might have provided to students in interpreting or updating or expanding on Aratus, would have reflected the grammarian's own limited or flawed understanding of the subject, so the depth and quality of astronomical knowledge spread through the occasional study of Aratus in grammar school might not have been universally good. It might not have been universal, either. Though the evidence shows it was not rare to find Aratus in grammar schools, the same evidence suggests it might not have been especially common.[120] Thus, the state of astronomical education in standard secondary education was perhaps, at best, a little better than negligible. And this pertains only to rudimentary astronomy. There are no comparable exceptions for any other scientific subject. For instance, an early 2nd century A.D. poem *Description of the Inhabited World* was basically a versified epitome by Dionysius of Eratosthenes' geography, and thus comparable to Aratus: aimed at school children and scientifically both light and obsolete.[121] But unlike the *Phenomena*, the *Description* is not well attested, and thus must not have been widely used.

Of course, as will be explained below, all this surviving evidence may be biased by excluding elite urban schools. Such schools would have been numerically exceptional, but may have served more students (as city populations far exceeded those of towns and rural districts). So science content in ancient secondary schools may have been more present than extant evidence confirms. This is particularly significant as scientists themselves are far more likely to have come from elite urban schools. Moreover, *outside* the context of schools, philosophical and scientific poetry was not uncommon, and thus a preparation for tackling poetry, particularly any scientific poetry (like the *Phenomena*) at lower levels of study would have equipped an interested student to approach scientific poetry independently or at higher levels of education, providing greater access to what we might call a "gateway drug" to further scientific study—for such poetry often had an implied purpose (sometimes even explicitly declared) of getting the

---

120. Morgan 1998: 43, although Morgan's source (Haarhoff 1920) is obsolete and pertains principally to the wrong period and place.

121. *OCD* 461 (s.v. "Dionysius (9) 'Periegetes'") and *EANS* 261–62 (s.v. "Dionusios of Alexandria, Periegetes"). This may be the same Dionysius who wrote *Gem Lore* and *Bird Lore* (among other works, cf. *EANS* 263–64, s.v. "Dionusios of Philadelpheia" and 259, s.v. "Dionusios (Lithika)," etc. *passim*), which may have been similar attempts at 'scientific' poetry.

reader more interested in its specific subject or in science and philosophy generally.[122]

Be that as it may, as already noted it was still not common for a member of the elite to advance beyond secondary education, even for men, and exceptional for a woman. These disparities became greatly magnified the further down one fell on the social ladder, so it is safe to say that most Romans who received any education went no further than grammar school. And as we have seen, that would have exposed them to little in the way of science or natural philosophy, and what they were exposed to was probably not of any consistent depth or quality. In terms of social values, this meant the standard "cultural package" that students, parents and communities expected a standard education to impart did not include science or natural philosophy—or even much in the way of philosophy at all beyond rudimentary ethics. Instead, its emphasis was a universalized form of cultural literacy: not merely the ability to read, write, and reckon, but the mastery of a particular vocabulary, not just of words, but of verses and allusions, and some common skills of aesthetic and technical literary analysis, which together allowed the educated to communicate with each other more adeptly, and share a common background of stories and interests, that separated them from the illiterate in often subtle and sometimes blatant ways. Getting ahead in society or simply maintaining one's status depended on this, on becoming 'one of us' instead of 'one of them', on being able merely to converse like an educated person.[123] Thus, to know your Homer, to employ and discuss principles of meter, to speak with an educated vocabulary, to use and understand obscure allusions to the characters and situations of classical theater and poetry, and similar abilities, comprised the most broadly useful package of skills, hence it became the package most

---

122. On all these points see Taub 2008. "Scientific" poetry of this kind included the *Aetna* (on volcanology), as well as Oppian's *Fishing* (on ichthyology), Manilius's *Astronomica* (on astrology, astronomy, and cosmology), and Lucretius's *On the Nature of Things* (on natural philosophy in general). Even Virgil's *Georgics* incorporated the scientific material on meteorology from Ps.-Theophrastus, *On Weather Signs*. And related to this genre is "technical" poetry, which Oppian's *Fishing* crosses into, plus such works as Boios's *On Raising Birds* and Eumelos's *On Raising Cattle*, which would also have versified both technical and scientific knowledge.

123. On this aim of using of education to separate the elite from the *hoi polloi*, see Whitmarsh 2001: 96–108.

broadly sought and taught. Science did not serve so broad a purpose, and was also generally more difficult to learn and understand.

Romans who completed only grammar school did have some of the skills in place to expand their scientific knowledge on their own if they wished, although there is no evidence of such a passion being common—and when present, it probably led to higher education anyway. Outside of such a formal setting, and the obscuring fog of oral lore and hearsay, there were only two ways to gain scientific knowledge: books and lectures. Harris observes that Varro's "head shepherd (*magister pecoris*) is supposed to be literate enough to use a book about ovine medicine [the medical care of sheep] which he carries around with him," which is an example of how a grammar student could informally expand his scientific knowledge through the use of books.[124] And in major cities, occasional public orations and lectures on scientific subjects would have afforded similar opportunities for a semi-educated person to expand their knowledge. But these options were not as convenient as they are now.

Books, for example, were nowhere near as available in antiquity as they are today. In the Roman period a single roll of papyrus, which held the equivalent of a single chapter in a modern book, cost on average four or five drachmas, the ancient equivalent of almost two weeks wages. So just the papyrus needed to produce a short five-chapter book would cost the equivalent of over two thousand dollars in modern U.S. currency.[125] And

---

124. Harris 1989: 256; referencing Varro, *On Agricultural Matters* 2.2.20.

125. Harris 1989: 195, 224–25 (corroborated by Hezser 2001: 145–46; whereas Winsbury 2009: 19–23 greatly underestimates this cost). Four to five drachmas equals 24 to 30 obols. The ancient equivalent of a 'minimum wage' was three obols per day (more or less—there was no fixed standard, cf. *OCD* 1567, s.v. "wages"). As of 2009 the federal hourly minimum wage in the U.S. was $7.25 and the standard full-time work-day consisted of eight hours, for $58 per day. So the modern social equivalent of one obol is in the vicinity of $19. Four or five drachmas thus approximates the value that $450 to $570 would have had to the average U.S. household in 2009, which multiplied by five makes $2250 to $2850, which is well over $2000. Books in codex form were less expensive, but not by enough to make much difference to the present point. Skeat 1982 argues a cost savings of 26%, and though many elements of his estimates and math are questionable (e.g. he greatly underestimates the number of lines that fit in a standard roll), even granting his conclusion would entail a $2000 book could be got for around $1500, hardly a discount of use to the average citizen. Moreover, Skeat fails to count the added

that was in Egypt where such rolls were manufactured. The costs of export would have substantially increased the price of rolls elsewhere, and in either case there was the additional cost of ink and the labor to hand copy every word (not to mention any markup for profit), which could bring the price of a book to five thousand dollars or more.[126] Rowland and Howe estimate that a ten-volume edition of *On Architecture* by Vitruvius would have cost 100 denarii, close to 100 drachmas, equivalent to more than $8,000.[127] Cribiore cites one occasion where 100 drachmas was paid *merely as a deposit* on the production of a book collated from multiple exemplars.[128] If producing your own copy, the usual practice was not to hire a copyist, but to rely on a specialized slave in-house, representing a large initial investment in the slave rather than a book-by-book expense. But only the very wealthy would have such slaves on staff, and their time would always be finite.[129] On the other hand, though used books could sometimes be found at a discount, their availability was nothing comparable to today, nor were their discount prices much more within reach of the poor.[130] The cheapest sale on record (and that possibly a mere fiction) was a bundle of six or so books "absolutely hideous in both condition and appearance" that Aulus Gellius

---

expense of binding the codex (whereas this cost is already included in the cost of papyrus rolls, which came pre-bound), so even on his own assumptions his estimated discount is too high (since binding books requires professional skill, as well as considerable time and a variety of materials). And *fine* codices even cost more than scrolls (Nicholls 2010).

126. See evidence and sources in Millard 2000: 165, who estimates that copying cost "six to ten" drachmas per roll, which is 30 to 50 drachmas (120 to 300 obols) for a five-chapter book, adding as much as $2000 to $6000. Thus even a small book could cost the equivalent of $4000 to $8000. See also Richards 2004: 165–70 who corroborates Millard; and see Cribiore 2001: 146–59 for a detailed discussion of the cost of books vs. more casual writing materials (such as for making notes and sales receipts). On other issues pertaining to the cost and difficulty of procuring books see Marshall 1976, Oleson 2008: 715–39, White 2009, Winsbury 2009, and Bagnall 2009: 256–81.

127. Rowland & Howe 1999: 1.

128. Cribiore 2001: 146–47.

129. Ibid. See also Cribiore 1996.

130. Starr 1990. For overall context see *OCD* 239–43 (s.v. "books, Greek and Roman," "books, poetic," and "books, sacred and cultic").

claims he found for sale on the docks, which he managed to get for "a few bronze," maybe a hundred dollars in modern currency.[131] But he considered this price "amazing and unexpected" even for books in the worst possible condition. In fact, from his description they probably contained much that was lost or illegible, hence their incredible cheapness probably reflected the fact that they were all but worthless on any real book market. And if only the bundle was being sold, not the individual books, then even an "amazing" price like this was pretty steep for the average Roman. Buying books was thus beyond the means of most, which meant they either had to borrow books from wealthy patrons or read them in public libraries, which were available in most major cities, a fact I'll discuss in chapter eight. But even if a literate layperson could thus get ahold of a book to read it, scientific works would still be almost inaccessible to the average reader, since most science texts required an understanding of specialized vocabularies or advanced mathematics. So without a more advanced education, access to scientific knowledge was still hindered or limited, even to those who might have an active interest.

Public speeches would have been more accessible to the semi-educated in this respect. The immense popularity of public rhetorical displays and 'show lectures' created opportunities for the casual public to hear and learn things, opportunities that did not exist on a comparable scale in any other ancient culture. This craze for public speaking was especially prominent in the early Roman empire, so much so the era is defined by it and commonly called the 'Second Sophistic'.[132] The Middle Ages almost produced something similar in the form of the sermon, although as a rule religious (and not scientific or any other) instruction far predominated there. For example, we have one sermon from Rabanus Maurus in the 9th century that teaches the basic science of eclipses, with the aim of dispelling superstitions in his

---

131. Aulus Gellius, *Attic Nights* 9.4.1–6. Assuming an average of six books of five rolls each, or thirty rolls, and assuming "a few bronze" as something in the vicinity of half a drachma, that would equal about $110, for an amazing discount of one sixtieth the cost of the papyrus alone. Though some scholars doubt the account of this sale, even as fiction the prices, condition, and buyer's reaction were probably realistic (as most good fiction aims to be).

132. On which see Bowersock 1974 and 2002, Anderson 1993, Whitmarsh 2005 and 2013, Schmidt & Fleury 2011, and *OCD* 1337–38 (s.v. "Second Sophistic"). See von Staden 1995 and Brunt 1994 for it's relevance to Roman science.

congregation that annoyed him, but this was very exceptional.[133] In the Roman era, it would not have been.

Though the content of most of these speeches in antiquity was literary, or at best historical or didactic, rather than scientific or philosophical, there was more than enough of the latter to be of use to an interested urbanite.[134] Galen, for example, describes an almost daily practice in Rome of scholars and other interested people gathering in the Temple of Peace to debate and chat specifically about science, where there also happened to be a public library that included scientific literature. The subjects discussed were not just medical as some scholars have implied, for Galen says all "theoretical sciences" (*logikas technas*) were talked about there.[135] Galen's writings contain numerous references to public medical lectures and anatomical demonstrations in cities throughout the Roman empire, giving the definite impression that this was a common phenomenon.[136] There is no reason to think other sciences did not receive similar treatment. As Dio Chrysostom said early in the 2nd century A.D., the subjects people loved to hear public lectures on included astronomy, meteorology, oceanography, physiology, anatomy, and other subjects in natural philosophy.[137] The Roman gymnasium in Delphi apparently had a resident astronomy professor who lectured on the subject for students and public alike, and that was not likely to be unique.[138]

---

133. Rabanus Maurus, *Homilies* 42.

134. Mudry 1986 (and evidence following here and in chapter seven). On the full range of popular sources of oral (and visual) education (mostly political, mythical, and religious in content) see Meggitt 2010: 56–61, 68–70 (with Toner 2010).

135. Galen, *On My Own Books* Kühn 19.19, 19.21; see also the (possibly) related remark in Galen, *On Venesection against the Erasistrateans at Rome* Kühn 11.194. On the library there, built by Vespasian (in 75 A.D.): Staikos 2000: 111. This library was accidentally destroyed by fire later in Galen's life (along with his personal collection of books and notes: Galen, *On Conducting Anatomical Investigations* 11.12; and now *On Avoiding Distress*, on which: Nicholl 2011, Jones 2009, Tucci 2008, and Mattern 2013: 259–77), which would have been around the dawn of the 3rd century. Whether it was restored is unknown.

136. For sources and scholarship on this see related note in chapter seven, with Tountas 2009.

137. Dio Chrysostom, *Discourses* 33.4–6.

138. König 2005: 50.

We know some people used the phenomenon of public lectures as a sort of free education.[139] Even Galen had some illiterate and poorly educated people attending his lectures.[140] Open lectures and oratorical displays like this would have been common in cities and could have disseminated some knowledge of natural philosophy, insofar as such information happened to be employed or mentioned in passing, or on the occasions when it was an actual subject of discourse. In addition to the phenomenon of public oration was the rarer occasion of hearing cases in court. Insofar as geometry and other scientific knowledge occasionally came up in the courts, and juries were drawn from the people, educated or not, jury duty may have occasionally exposed some Romans of all educational levels to a smattering of medicine, mathematics, and other science.[141] Since such cases would turn on the point at issue, an advocate must have had to explain the relevant math or science to the jury, insofar as the case required. But since the ancient legal system made little significant use of scientific evidence or testimony, occasions for it would have been rare, and its content very limited. But it would not be insignificant. According to Vivian Nutton, "doctors might be called upon to testify," as for example in Roman Egypt, where "medical certification in cases of wounding or suspicious death" was "apparently a common procedure" attested in papyri.[142] This was surely as much the case in other provinces. As we'll see in a moment, medical science was certainly expected to arise in court according to various rhetorical exercises.

Whether through public orations or jury trials, this ubiquity of public discourse would have produced a chain of lore and hearsay, as facts heard and learned by audience members came to be repeated and disseminated among their acquaintances, and from them to others, and so on, ultimately filtering

139. Cribiore 2001: 58.

140. According to *On the Affections and Errors of the Soul* 2.2 (= Kühn 5.64–66).

141. Juries drawn from the people: Quintilian, *Education in Oratory* 12.10.53. Geometry and science in court: Quintilian, *Education in Oratory* 1.10.36 (see also Cuomo 2001: 4–5, 20–21, 215–17 and Campbell 2000: xxix, xxxvi).

142. Cf. *OCD* 1089–90 (s.v. "pathology"). On medicine as an issue in court see Amundsen 1978 and 1979, who claims the legal system in Roman Egypt may have been unique in employing doctors forensically (as he proves it did do), but he presents no evidence it was in fact unique there, nor is there any reason to believe it would be.

such information down into the pool of general knowledge. However, the quality of scientific knowledge that survived this process would not have been particularly profound, consistent, or good, and would have mingled indiscriminately with nonsense and poppycock. At best, we can expect that due to this unusual practice of members of the educated elite delivering long eloquent speeches on a regular basis to large crowds, the amount and quality of scientific knowledge that entered general public consciousness would at least have been greater than in other ancient cultures—or indeed the Middle Ages. But that's only by a relative measure. In absolute terms, the significance of this phenomenon could not have been great. In the end, the only real exposure Romans ever likely got to good science content was in higher education—with one notable exception...

5. The Enkyklios Paideia

That exception was the so-called *enkyklios paideia*, a more-or-less standardized ideal of a 'well rounded' education. This was developed and most widely embraced among the Greeks, but by the early Roman empire it was also passionately advocated among the best Latin educators. It was no doubt widely realized—just not universally. The phrase *enkyklios paideia* literally means "curricular education" or in effect "well-rounded," "regular," or "general education." The seven most popular disciplines of the *enkyklios* (grammar, rhetoric, dialectic, arithmetic, geometry, music, and astronomy) were described in antiquity as 'the liberal arts' because they were intellectual skills pursued by free men who had leisure to study them. Hence they became symbols of status, a way to distinguish a free man of means from slaves and the working class, who (it was usually assumed) did not have the time or means to range so widely or deeply in their educational studies.

In practice, only the best students completed the entire course. Even so, those teachers of rhetoric who upheld the highest standards of their day expected students to have completed the full *enkyklios*, as did many more professors of philosophy, and we can be sure so did many doctors, engineers, and astronomers. There is ample evidence that many among the educated elite did complete this standard preparatory curriculum, in all periods, from the Classical to the early Roman. Hence the full curriculum was only 'idealistic' in the sense that it was never realized universally, although the idle wish that it would be was often voiced. Strabo even declares that those who have not completed a full course in the liberal arts will be incapable of sound judgement in almost every important matter, a view that would condemn almost everyone he knew—unless he knew many peers who had completed the course, as he himself must have.[143] Nevertheless, as Clarke says, except for grammar and rhetoric, the full seven arts "were not pursued (except by a few specialists) beyond a quite humble level." Several ancient writers kept arguing how students should pursue them more than they do, or complained that many did not pursue them at all. So the reality must have been that most people "thought it unnecessary to study anything but grammar and rhetoric" even while a sizable minority continued to master the full curriculum, though most only superficially.[144]

The complete *enkyklios paideia* can be loosely equated with modern 'college prep' courses, which most students at the secondary level do not take, but those who do take them improve their chances of getting into the best universities. Likewise, in antiquity most students at the secondary level did not complete the *enkyklios*, but many did, and by doing so they improved their chances of being accepted into the tutelage of the best professors. That was not the only motive for completing this special curriculum, however. Parents might often have expected it of their children, simply to ensure they were adequately cultured and thus capable of maintaining their elite status in society. The education of Cornelia, for example, clearly involved completion of several subjects in the *enkyklios*, and probably reflects a trend among select parents seeking to provide a complete education for their daughter in order to improve her prospects for a good marriage. They certainly did not

143. Strabo, *Geography* 1.1.22.

144. On all these points, see Clarke 1971: 6–7. As witnesses Clark references Quintilian (late 1st century, early 2nd century A.D.), Dionysius of Halicarnassus (late 1st century B.C.), Aelius Theon of Alexandria (1st century A.D.), Lucian of Samosata (2nd century A.D.), and Galen (late 2nd century A.D.), to which we can add the remarks of Aulus Gellius (*Attic Nights* 1.9.6–7). Galen often complains of ignorance of advanced mathematics among his peers (e.g. Iskandar 1988: 158, §P.68,14–15, though one could just as easily remark upon the same popular ignorance of geometry and trigonometry today, the very subjects Galen means). Many students nevertheless did study the full curriculum: Clarke 1971: 7, 111; Bonner 1977: 78; Marrou 1964: 265–79, 372–73 (cf. 1956: 183–84); Rawson 1985: 4–5 (and examples below). For a partial survey of some of the scientific content of this curriculum see Marrou 1964: 265–79, 372–73, 410. For general discussion of the content and status of the *enkyklios* (especially in the Roman period) see Stahl 1971: 90–99; Bonner 1977: 77; Rawson 1985: 156–57; J. Barnes 1988; Russell 1989; and Doody 2009.

undertake the great expense in time and money such education required in order to groom her for a career. Both boys and girls could receive such attention for much the same reason—wealthy parents did not want their kids to look like hicks or hacks in high company. Thus, though in practice the *enkyklios* formed a special preparatory curriculum at the secondary level that would be completed by the best students who advanced to any form of higher education (and especially those with an eye on studying philosophy or any actual science), it was also for some an end-in-itself, a way of seeming 'more educated' without aiming at any particular occupation.

Though all contemporary historians of ancient education agree that completion of the *enkyklios* was not standard or common, it cannot have been as rare as some have implied. Since so many ancient authors clearly had completed it (some even specifically said they did or assumed many others had), it clearly was pursued more than idiosyncratically, and thus instruction in it had to have been widely available, at least to urban populations. This is significant because the *enkyklios* contained a substantial amount of science and science-related content. Thus, those students at the secondary level who were exposed to it would have had a better science education than even many who completed a full course of higher education in rhetoric (as we shall see below), *and* would have had better prospects of entering a course of higher education (such as in philosophy or medicine) that surveyed even more science content, *and* would be far better prepared to expand their scientific knowledge on their own through reading and attending lectures, since they will have received much of the basic foundations for further advanced study.

The standard *enkyklios paideia* was later divided in the Middle Ages into the 'trivium' ("the threefold study") and the 'quadrivium' ("the fourfold study"), but though this terminology came late, it reflected a reality that was already well-established by the Roman period, namely that most students of rhetoric completed only the first course while only the best or most fortunate students completed the second as well.[145] The first course, the trivium,

145. See Stahl 1962 and 1971, who documents a decline in the quality of this education in Latin schools during and beyond the 4th century, although he over-exaggerates the quality of this education *before* the Roman period. As we shall see in a moment, however, Stahl's claim that "the only people who seriously promoted the study of all seven liberal arts were philosophers" is false (Stahl 1971:91). Most philosophers did have a special interest in the mathematical and scientific content

consisted of grammar, rhetoric, and dialectic (i.e. logic), generally studied in that order, with some gradual overlap at each juncture. Grammar would be studied at the secondary level, plus occasionally some preparatory studies related to rhetoric, while most rhetoric and dialectic would be pursued only at the next and often final stage of educational advancement. In the end, to complete a full course of higher education in rhetoric was in effect to complete the trivium. We will thus address the science content of the trivium in the next chapter, when we come to higher education. The second course, however, was highly science-oriented. Later called the quadrivium, this consisted of arithmetic, geometry, music, and astronomy.[146] The ancients were aware of the fact that this excluded the two other advanced sciences in antiquity, medicine and engineering, and some educational idealists promoted the idea of including them in the curriculum, but there is no evidence they ever succeeded.[147] As we'll see later, many authors clearly

---

of the *enkyklios*, but so did the most noteworthy professors of rhetoric.

146. Even before the terminology existed, the trivium had become the standard universal education and was long considered a complete literate education, while the quadrivium was explicitly identified as a complete mathematical education at least by the end of the 1st century A.D., when Nicomachus of Gerasa wrote in his mathematics textbook, *Introduction to Arithmetic* 1.3.4, that "the four" subjects included under "mathematics" were arithmetic, geometry, harmonics (i.e. music), and astronomy. For background see *OCD* 1014 (s.v. "Nicomachus (3)"), *DSB* 10.112–14 (s.v. "Nicomachus of Gerasa"), and *EANS* 579 (s.v. "Nikomakhos of Gerasa"). A Latin edition of this was produced by Apuleius in the 2nd century (cf. S.J. Harrison 2000: 31–32). Nicomachus also produced textbooks on harmonics and geometry and it is reasonable to assume he had planned or even completed one on astronomy (for evidence see D'Ooge et al. 1926: 81–82). But already in the early 4th century B.C., Archytas (a Pythagorean expert in harmonics) asserted that arithmetic, geometry, music, and astronomy are all interrelated, so some idea of the quadrivium had appeared even well before Roman times (*DSB* 1.231–33, s.v. "Archytas of Tarentum," *OCD* 145, s.v. "Archytas," and *EANS* 161–62, s.v. "Arkhutas of Taras").

147. On efforts to classify medicine as a liberal art see Kudlien 1976, but the idea is most eloquently voiced in Plutarch, *Advice on Keeping Well* 1 (= *Moralia* 122d-e). In contrast, gymnastics and drawing fared better. Aristotle had said in his day there were four customary subjects of education—reading and writing, gymnastics, music, and drawing (Aristotle, *Politics* 8.2.1337b)—and several Roman authors agreed drawing should be included in every student's education (Pliny the Elder, *Natural History* 35.36.77; Plutarch, *Life of Aemilius Paullus* 6.5; and probably

had knowledge of these subjects, however, so their pursuit must have been possible for laypersons, even if only rarely or imperfectly.

In general, at this level arithmetic went beyond mere rudimentary numeracy and included more advanced skills of adding, subtracting, dividing and multiplying, and established a greater competency to deal with numbers and basic mathematics—and at more advanced levels it included the ancient equivalent of algebra, in the same way advanced geometry included trigonometry.[148] Ancient arithmetic also included combinatorics and logistics and other arts, while ancient geometry included many other subfields including isoperimetry.[149] Though these studies did not involve

---

Varro, cf. Rawson 1985: 193, 198) while gymnastics definitely won a place in standard education, though perhaps more so in Greek communities than Latin: Galen is describing a Greek's education in *Advice for an Epileptic Boy* 2–5, which simply assumes a student went to gym class, whereas we find some distaste for gymnastics in the Latin author Quintilian (as we'll soon see). On the issue of athletics in ancient education in general see König 2005 and Petermandl 2014.

148. Ancient algebra is discussed in Christianidis & Oaks 2013 and Derbyshire 2006:31–42; see also *DSB* 4.110–19 and 15.118–22 (s.v. "Diophantus of Alexandria"), *EANS* 267–68 (s.v. "Diophantos of Alexandria"), and *OCD* 465 (s.v. "Diophantus"), which correctly dates him "between 150 BC and AD 280," hence probably Roman-era. A good case for dating Diophantus to the 1st century A.D. is presented in Knorr 1993 and Russo 2003: 322–23 (esp. n. 230). Some basic principles of algebra might date as far back as the 4th century B.C., cf. *DSB* 13.399–400 (s.v. "Thymaridas") and *EANS* 808–09 (s.v. "Thumaridas (of Paros?)"). Similarly, while basic principles of trigonometry were already developed as early as the 3rd century B.C., plane and spherical trigonometry were fully formalized by Menelaus in the 1st century A.D., cf. *DSB* 9.296–302 and 15.420–21 (s.v. "Menelaus of Alexandria"), *EANS* 546 (s.v. "Menelaos of Alexandria"), and *OCD* 932 (s.v. "Menelaus (3)"), as well as *OCD* 1507 (s.v. "trigonometry"), with analysis in Russo 2003: 52–55 and Van Brummelen 2009 and 2013. Modern systems of trigonometry and algebra are entirely different, as both sciences were all but forgotten and had to be reinvented, this time by medieval Indians and Muslims respectively, who improved both before diffusing them to the West. But the ancient systems still worked and achieved the same basic goals.

149. Hein 2012. On ancient combinatorics see: *DSB* 15.220 (in s.v. "Hipparchus") with Russo 2003: 281–82, Netz 2003: 283–84, Netz, Acerbi & Wilson 2004, Netz & Noel 2007: 54–59, 233–60, Bobzien 2011, and the bibliography in *DSB* 15.223–24 (even Plutarch was aware of combinatorics: Plutarch, *Tabletalk* 8.9 = *Moralia* 732f-733a). On ancient logistics and other mathematical fields see: Geminus (1st century B.C./A.D.) as paraphrased in Proclus (5th century A.D.), *Commentary on*

exposure to natural philosophy as such, it provided the basic foundation for mathematical reasoning and comprehension, which was essential to the sciences (even, to a lesser extent, the life sciences). Geometry was even more important in this regard, since geometrical facts and analyses were essential to the physical sciences, while quasi-geometrical methods of proof were adopted even by the best practitioners of the life sciences. Geometry also involved exposure to astronomy, geography, optics, and mechanics as examples for study and application.[150] Music, meanwhile, did include the actual art and practice of music, but it always included music theory as well, almost by necessity—since only by linking its practice to a rational theory could music be advanced as respectable for cultured ladies and gentlemen.[151] This meant that the study of music in the *enkyklios* actually involved some study of the science of harmonics, which encompassed both music theory and basic acoustics. Finally, the science content of astronomy included time keeping and calendrics; identifying stars and constellations; principles for measuring and observing the heavens year round, and predicting where stars, sun, and planets will be; some meteorology; and basic astrophysics (e.g. causes of lunar and solar eclipses; sphericity of the earth; names,

---

the First Book of Euclid's 'Elements' pr.1.13.38–42, with translation and commentary in Evans & Berggren 2006: 43–48, 243–49. On isoperimetry see: *DSB* 14.603–05 (s.v."Zenodorus"), *OCD* 1588 (id.), and *EANS* 845 (s.v."Zenodoros"). The educator Quintilian shows a sound grasp of the uses and principles of isoperimetry and gives several examples of why generals, historians, surveyors, and lawyers needed to learn it (*Education in Oratory* 1.10.39–45). Examples of its application and discussion are found in extant surveying manuals from the early Roman empire (e.g. Campbell 2000: 12–13) and it found use even in biology (e.g. see Cuomo 2000: 57–90 for its use in apiology; Aristotle, *Posterior Analytics* 1.13.79a for its use in medical physiology). So clearly these subjects were often known and even learned by non-specialists.

150. For example, Vitruvius, *On Architecture* 1.1.4, 1.1.7, 1.1.16.

151. Morgan 1998: 35; cf. Vitruvius, *On Architecture* 1.1.8–9. That many among the elite had received such an education in the science (and not merely the craft) of music is shown in Barker 1994: 59–60, and (though less thoroughly) in Vendries 2004; note that Vendries incorrectly believes there is no evidence "d'une anticipation du *trivium*...et du *quadrivium*" ("of an anticipation of the *trivium*...and the *quadrivium*") in the early Roman period, a conclusion refuted by evidence in our present chapter, as well as by the survey in Stückelberger 1965: 32–44, 46–52 and comments in J. Barnes 1988: 56–57.

nature, and courses of the planets).[152] All of these studies provided a certain inoculation against formal and informal 'skepticism' and served as a model for scientific argument, proving the power of linking empirical observation with mathematical description, as Galen would so forcefully argue in the second book of his *On the Affections and Errors of the Soul*, and repeatedly elsewhere.

In the end, though these studies would only *introduce* a student to the best achievements of ancient science, and though they did not always reach into the deeper issues of natural philosophy, they provided essential and valuable foundations for the further study of every advanced science of the day. They were also a likely stimulus motivating more students to enter and advance the sciences. Though many no doubt had their eye on the sciences merely for the career prospects they offered, as more students were exposed to the science content of the *enkyklios* more would have been ensnared by any innate passion for science they may have discovered in themselves. Hence attention to the *enkyklios* must have increased the number of students entering advanced scientific studies out of a passion for knowledge, rather than only for more material aims, and this meant the educational ideal of the *enkyklios*, though never universally realized, was still a valuable asset to the progress of ancient science generally. And more importantly for our concern, the fact that the *enkyklios* was studied widely enough not to be regarded as a fringe pursuit, and was even more widely praised than studied, means there was a significant portion of the educated populace that held science and natural philosophy in high esteem—a much larger and more vocal contingent than in any other ancient culture. To what extent this was comparably the case in the Middle Ages thus needs to be explored.

This complicates the picture of ancient education that we have generally seen so far. Though the place of science in ancient education was minimal *overall*, there was nevertheless a significant portion of educated society, and of the educational system supporting it, embracing a more extensive exposure to the sciences, enough to distinguish ancient Greco-Roman culture. The presence and status of the quadrivium in ancient education also demonstrates that even from the very beginning the most widespread

152. On the actual content of astronomy taught in the encyclical curriculum see Evans & Berggren 2006: 8–12 (and for a textbook applying astronomical science to the philosophy of cosmology around the same time see Bowen & Todd 2004).

attention ever paid to any scientific subjects in ancient secondary education emphasized fully mathematized sciences, which were likewise empirical (harmonics relying on demonstrations with instruments, astronomy on observation, and geometry on practical applications as proofs of concept, from such diverse fields as optics and mechanics, and all involving reference to scientific instruments of diverse kinds).

While the *enkyklios* probably represented the best education that the luckiest kids received, a truly exceptional education probably touched on the remaining two most advanced sciences of the ancient world as well: medicine and engineering. These sciences should be understood in their broadest possible sense: medicine included anatomy, physiology, and pharmacology, and encompassed zoology, botany, and mineralogy; engineering included physics, astronomy, cartography, optics, harmonics, and mechanics, in addition to architecture (chemistry never received a comparable scientific footing in antiquity—or in the Middle Ages). In the middle of the 1st century B.C., Varro wrote an encyclopedia called the *Disciplines*, which introduced the whole gamut of Greek education into Latin, in nine books—one each on the seven disciplines of the *enkyklios paideia* and one each on medicine and engineering.[153] Though the entire work is lost, from numerous references and quotations in other authors we know Varro included under the study of geometry not only the abstract mathematical discipline (summarizing Euclid, for example) but also the empirical disciplines of optics, harmonics, astronomy, surveying, and geography, all involving extensive practical applications, demonstrating that considerable scientific content could be taught under the rubric of 'geometry'.[154]

---

153. Known in Latin as the *Disciplinae* or the *Disciplinarum Libri IX*. Although this appears to have been the first such book in Latin, it was certainly not the first time Romans were exposed to these subjects, since their bilingual elite had already been familiar with Greek education—many had even studied in Greece—for a century or more before Varro wrote. See Stahl 1971: 96 (and 7: n. 11); Clarke 1971: 2; and *DSB* 13.588–89 (s.v. "Varro, Marcus Terentius"), *OCD* 1441 (s.v. "Terentius Varro, Marcus"), and *EANS* 774–78 (s.v. "M. Terentius Varro of Reate"). A Latin epitome (or inferior plagiarization) of Varro's encyclopedia may have been produced in the mid-3rd century by Censorinus, of which fragments survive (see *DSB* 3.175–76, s.v. "Censorinus," *OCD* 296, id., and *EANS* 212, s.v. "Censorinus (II)").

154. Stahl 1971: 44–53; Rawson 1985: 158–59.

Although Varro included medicine and engineering in his survey of educational topics, it is unlikely a significant number of people in any generation actually undertook all nine fields of study, even to dabble in them. Some few, however, may have gotten as far as dabbling in the full nine.[155] The poet Virgil was said to have studied both medical and mathematical sciences.[156] And the doctor Galen demonstrates throughout his writings that he had studied all nine disciplines more than trivially, exhibiting his own mastery of rhetoric while writing sophisticated books in grammatical science and formal logic, as well as of course medicine, and at the same time frequently exhibiting competence in all the mathematical sciences, even promoting engineering in place of rhetoric as a subject to be included in the *enkyklios*.[157] But there is no ancient scientist on record who mastered both medicine and engineering as a field *of research*. Though some, like Galen, no doubt knew a lot from both fields, the practice and education required to master one were apparently too great to leave any significant time for the other, so ancient scientists came to be divided into the physical and life sciences, according to whether they pursued engineering or medicine as their primary interest. If this held even for the greatest geniuses of the age, it

155. Evidence of considerable knowledge and interest in medical science among educated laypeople in the Roman period is surveyed in Nutton 1985 and 2004: 252–53 (with Ballér 1992 and Durling 1995). Note that an encyclopedia of the arts superior to Varro's was produced a century later by Aulus Cornelius Celsus, which also included medicine as a subject. We're not sure of the full range of subjects this treated (we have only scattered hints in Quintilian, *Education in Oratory* 12.11.24, and Columella, *On Agricultural Matters* 1.1.14), but its treatment of medicine is rather superb: see *DSB* 3.174–75 (s.v. "Celsus, Aulus Cornelius"), *NDSB* 2.81–84 (s.v. "Celsus, Cornelius (Aulus)"), *OCD* 377 (s.v. "Cornelius Celsus, Aulus"), *EANS* 217–19 (s.v. "A. Cornelius Celsus"), with Scarborough 1970: 298–302. For a brief comparative analysis of the encyclopedic works of Cato, Varro, Celsus, and Pliny, see Doody 2009.

156. Suetonius, *Virgil* 15, says that in his early education Virgil devoted himself to "the study of medicine and to a great extent mathematics."

157. Galen's education will be discussed in chapter seven. But for some examples of Galen's inclusion of engineering in an ideal education and of his own considerable knowledge of the subject see Galen, *On the Affections and Errors of the Soul* 2.2–5 (= Kühn 5.64, 5.68–5.69, 5.80–5.91), though more examples will be cited later. Galen's effort to promote a full encyclical education is also reflected in his treatise *Exhortation to Study the Arts*.

surely held even more for everyone else. Of course, this remains true today. Rarely (if ever) do we see any scientist mastering both a life science and a physical science, much less producing research in both.

The restriction of the scientific content of the *enkyklios* to two specific fields (music and astronomy) and one general field (geometry; arithmetic touched on little in the way of science) derives in part from the influence of Plato and other sophists of his generation centuries before the Roman era, as such men had a particular philosophical interest in these subjects, while engineering and medicine had not yet advanced far enough then to impress them in the same way (though they had by the time Varro wrote).[158] However, the endurance over centuries of the traditional studies probably reflects the fact that all four subjects of the quadrivium were related to each other by common mathematical content, and in practice may have amounted to the most the average good student could reasonably be expected to master without becoming a specialist.

Aristotle's observation on this point is probably just as applicable to the Roman period as his own: "young men (*neoi*) become geometers and mathematicians and wise in other subjects like that but do not, it is thought, become prudent (*phronimos*)," because prudence requires considerable experience, which young men lack. Thus, if "someone asks why a child (*pais*) might become a mathematician but not a sage (*sophos*) or a natural philosopher (*physikos*)," Aristotle answers that mathematics is learned solely "in the abstract" but wisdom and natural philosophy derive their principles "from experience" and therefore "young men (*neoi*) do not trust them, but merely repeat them," unlike mathematics, whose principles and conclusions can be seen directly, and therefore believed, the instant they are taught.[159] Thus, the simplest and most mathematical sciences were suitable "for boys," but fields as deeply dependent on extensive observation and experience as medicine and engineering would only be suitable for adults (which in antiquity could mean anyone over the age of fourteen). This may have been uncharitable to even an average child's actual capabilities, but if it was

158. Plato, *Republic* 7.525a-531e (and see Stahl 1971: 90–98). On the enormous advances in these sciences after the era of Plato and Aristotle, continuing into the Roman period, see Carrier 2010, Russo 2003, and Rihll 1999. I will treat the subject in some detail in *The Scientist in the Early Roman Empire*.

159. Aristotle, *Nicomachean Ethics* 6.8.1142a.

nevertheless a pervasive view it would explain why mathematical subjects came to the fore in the ancient curriculum.

Since medicine and engineering never won their way into any general scheme of education (and thus were studied almost exclusively by specialists and the most devoted of laypeople), we will continue to speak only about the standard content of the *enkyklios*, which at least reached a sizable minority of the educated elite. Moreover, since the trivium became the backbone of ancient secondary and higher education and yet contained little or no scientific content, we shall focus only on the quadrivium here. This section of the curriculum was often seen as preparatory for the study of philosophy, and was thus pursued mostly (though not only) by students aiming at or being groomed for a higher education in philosophy or science—and, we can assume, those who wished to move in such circles (or marry into them). Although not all philosophical schools required the quadrivium as preparation, the most popular (and most 'respectable') schools did—in other words, the Stoics, Platonists, and Aristotelians.[160] Plato, for example, argued that students should study astronomy and harmonics, which in turn required the study of geometry and arithmetic, because these will elevate the mind and foster a sense of community and kinship among students sharing the experience of learning them.[161]

As Marrou concludes, it was still the case that "the *grammatikos* very commonly took precedence over the *geometres*, and, apart from a few specialized vocations, such as architecture," or astrology, medicine, philosophy, "Greek culture of the Hellenistic and Roman periods was predominately literary at the expense of study of the sciences."[162] Marrou goes even further and supposes "philosophy never recruited more than a tiny minority of the élite minds," and therefore "statistically, so to speak, Isocrates decisively defeated Plato" in defining the dominant focus and course of

160. Bonner 1977: 78–79; Clarke 1971: 3–5.

161. Plato, *Republic* 7.525a-531d. Plato said the final field of study this would all lead to was dialectics (ibid. 531e-532d), so with rhetoric and grammar (in fact the latter being a necessary preparation for the former), that completes the seven-discipline curriculum that nearly everyone would accept as the ideal education for at least a thousand years.

162. Marrou 1981: 193.

education in antiquity.[163] Marrou here names Isocrates, a contemporary of Plato, on the supposition that Isocrates attacked Plato's educational program and proposed the alternative 'literary' program that came to be commonly pursued. This supposedly represents an age-old battle between philosophers and orators over control of the educational system of antiquity, and by virtue of their vastly superior numbers, and the greater evident utility of their art, the orators always won. But such a picture distorts the reality.[164]

In actual fact, all the leading advocates of making oratory the central focus of education—including Isocrates himself—actually embraced the mathematical and scientific content of Plato's program. The failure of such a broad educational curriculum to become universally adopted, remaining instead the privilege of an even more select few than received any education at all, reflects more the practical realities of the time than any victory of Isocrates over Plato. Plato's entire vision for society in the *Republic* was a fascist pipe-dream that never came anywhere near to being realized, not even in Plato's own school, which certainly only a select and privileged few in the Greek world ever had a chance of attending anyway. So whatever fantasies Plato may have had on the subject of education can hardly be taken as representative of what philosophers as a whole actually sought to achieve, much less elite society as a whole. But what Isocrates advocated was more realistic and more favorable to mathematical and scientific subjects than Marrou allows. Since Isocrates' opinions on this subject were still actually read and taught in schools throughout the Roman empire, and since Marrou believes (correctly) that these views came to define Roman education generally, we shall here pay close attention to them.

Isocrates was sometimes critical of the utility of science education, but even he succumbed to what Takis Poulakos calls the "impulse to appropriate rather than undermine existing educational practices" and hence:

> This impulse explains why, when [Isocrates] puts forth his views about the type of educational training he considers most appropriate for Athenian youth, he includes the study of subjects he had been most critical about. As he envisions it in the *Antidosis*, the ideal curriculum would include, for example, the study of the sciences, the very field he earlier condemned

163. Marrou 1981: 195.

164. As Demont 2004 shows with respect to the same argument in Marrou 1964.

> as irrelevant to "private or to public affairs" ([Isocrates, *Antidosis*] 262). For even though he considers instruction in astronomy and geometry to have no intrinsic value and no pertinence to students' "ability to speak and deliberate on affairs" ([Isocrates, *Antidosis*] 267), he recognizes the protreptic value of these studies and labels them "gymnastic of the mind" ([Isocrates, *Antidosis*] 266). Because such training would be more advanced and more helpful in disciplining and sharpening young minds ([Isocrates, *Antidosis*] 265) than the standardized, elementary instruction in grammar and music, he would place the study of astronomy and geometry at a level of instruction following the elementary level ([Isocrates, *Antidosis*] 262).[165]
>
> In fact:
>
> According to Isocrates, a training in sciences would improve students' aptitude for mastering greater and more serious studies, or, at the very least, it would keep their minds occupied: "even if this learning can accomplish no other good, at any rate it keeps the young out of many other things which are harmful" ([Isocrates] *Panathenaicus* 27).

Even though Isocrates wrote these things well before our period of concern, this text was still being widely copied, disseminated, and read in Roman times, and the works of Isocrates were among those commonly read even in schools. Thus, his opinions would still be influencing educators and students of the Roman period, and yet his opinions on this subject produced no recorded objections. It must also be observed that Isocrates is speaking of what was expected of *general* education, so even the ambivalence toward science that *is* found in Isocrates did not entail any objection to specialists pursuing astronomy or other sciences, or rejection of the value to society of the work of astronomers and other scientists. That it was useless for every citizen to learn astronomy as well as a specialist would was true then and remains true to this day. Isocrates saying so is just common sense. It is not an assault on any *realistic* alternative offered by Plato, since Plato's ideals were as unrealistic then as they would be now. Thus, though the educational values of Isocrates did win the day, this did not mean he or his adherents

165. Poulakos 1997: 98. Like Poulakos, Hutchinson 1988 also produces a more accurate analysis of how Plato, Isocrates, and Aristotle *really* differed (and as often agreed) on the purpose, process, and ideal content of education. Wareh 2012 argues that ongoing debates between Isocrateans and Platonists produced this alignment of interests.

believed that neither astronomy nor astronomers had any social or practical value. Hence we find no actual hostility, ambivalence, or negativity toward astronomy in Isocrates—just against the absurd notion that every student should 'become an astronomer'.

This is so evident from what Isocrates actually says that he must be quoted in full, especially since, as we shall see, this appears to have become the common opinion among Roman educators as well and reflects the pervading attitude of the time:

> I believe that the teachers who are skilled in disputation and those who are occupied with astronomy and geometry and studies of that sort do not injure but, on the contrary, benefit their pupils, not so much as they profess, but more than others give them credit for. Most men see in such studies nothing but empty talk and hair-splitting, for none of these disciplines has any useful application either to private or to public affairs; nay, they are not even remembered for any length of time after they are learned because they do not attend us through life nor do they lend aid in what we do, but are wholly divorced from our necessities. But I am neither of this opinion nor am I far removed from it; rather it seems to me both that those who hold that this training is of no use in practical life are right and that those who speak in praise of it also have truth on their side. If there is a contradiction in this statement, it is because these disciplines are different in their nature from the other studies which make up our education.
>
> For the other branches avail us only after we have gained a knowledge of them, whereas these studies can be of no benefit to us after we have mastered them unless we have elected to make our living from this source, and otherwise only help us while we are in the process of learning. For while we are occupied with the subtlety and exactness of astronomy and geometry and are forced to apply our minds to difficult problems, and are, in addition, being habituated to speak and apply ourselves to what is said and shown to us, and not to let our wits go wool-gathering, we gain the power, after being exercised and sharpened on these disciplines, of grasping and learning more easily and more quickly those subjects which are of more importance and of greater value. I do not, however, think it proper to apply the term "philosophy" to a training which is no help to us in the present either in our speech or in our actions, but rather I would call it a gymnastic of the mind and a preparation for philosophy. It is, to be sure, a study more advanced than that which boys in school pursue, but it is for the most part the same sort of thing.
>
> For they also, when they have labored through their lessons in

grammar, music, and the other branches, are not a whit advanced in their ability to speak and deliberate on affairs, but they have increased their aptitude for mastering greater and more serious studies. I would therefore advise young men to spend some time on these disciplines, but not to allow their minds to be dried up by these barren subtleties, nor to be stranded on the speculations of the ancient sophists, who maintain, some of them, that the sum of things is made up of infinite elements; Empedocles that it is made up of four, with strife and love operating among them; Ion, of not more than three; Alcmaeon, of only two; Parmenides and Melissus, of one; and Gorgias, of none at all. For I think that such curiosities of thought are on a par with jugglers' tricks, which, though they do not profit anyone, yet attract great crowds of the empty-minded, and I hold that men who want to do some good in the world must banish utterly from their interests all vain speculations and all activities which have no bearing on our lives.[166]

And:

[Yet] so far from scorning the education which was handed down by our ancestors, I even commend that which has been set up in our own day—I mean geometry, astronomy, and the so-called eristic dialogues, which our young men delight in more than they should, although all the older men declare them insufferable. Nevertheless, I urge those who are inclined towards these disciplines to work hard and apply themselves to all of them, saying that even if this learning can accomplish no other good, at any rate it keeps the young out of many other things which are harmful. Nay, I hold that, for those who are at this age, no more helpful or fitting an occupation can be found than the pursuit of these studies. But for those who are older and for those who have been admitted to man's estate I assert that these disciplines are no longer suitable. For I observe that some of those who have become so thoroughly versed in these studies as to instruct others in them fail to use opportunely the knowledge which they possess, while

166. Isocrates, *Antidosis* 261–69 (translations adapted from the Perseus Digital Library). Notably, what Isocrates is saying may be correct: the study of the sciences, even if the material learned is itself never used or retained, might nevertheless increase what is today called the Intelligence Quotient, or IQ, explaining what is now called the Flynn Effect: a documented rise in IQs among nations with strong national primary and secondary science education, in particular the form of intelligence known as Hypothetical-Categorical-Abstract Reasoning, which is of particular value to citizens in a democracy (in exactly the way Isocrates claimed) and to any culture that would become in any way scientific. See discussion and sources in Cheyne 2010.

> in the other activities of life they are less cultivated than their students—I hesitate to say less cultivated than their servants.
>
> I have the same fault to find also with those who are skilled in oratory and those who are distinguished for their writings and in general with all who have superior attainments in the arts, in the sciences, and in any specialized skill. For I know that the majority even of these men have not set their own house in order, that they are insupportable in their private intercourse, that they belittle the opinions of their fellow citizens, and that they are given over to many other grave offenses. So I do not think that even these may be said to partake of the state of culture of which I am speaking. Whom, then, do I call educated, since I exclude the arts and sciences and specialties? First, those who manage well the circumstances which they encounter day by day, and who possess a judgement which is accurate in meeting occasions as they arise and rarely misses the expedient course of action; next, those who are decent and honorable...agreeable and reasonable...who hold their pleasures always under control and are not unduly overcome by their misfortunes...and most important of all, those who are not spoiled by successes and do not desert their true selves and become arrogant.... Those who have a character which is in accord, not with *one* of these things, but with all of them—these, I contend, are wise and complete men, possessed of all the virtues.[167]

Almost exactly the same argument is found in the letters of Seneca, a Roman philosopher and statesman of the 1st century A.D., where he specifically summarizes and addresses the *enkyklios* in detail.[168] A similar sentiment is voiced by one of the characters in Cicero's dialogues, and Cicero was often held to be the quintessential Roman philosopher and statesman even long after his death in the 1st century B.C.[169] But the point to observe here is that, like Cicero and Seneca, Isocrates is not attacking all astronomy and astronomers nor all scientists and technicians as useless or uncouth. He is saying that *merely* being an astronomer or scientist or technician (or orator or poet or *any* sort of skilled person) does not make someone educated, for they must *also* be good and virtuous and reasonable people before Isocrates will consider them educated. Thus, given his remarks about what constitutes a good education (and his qualifying his own remarks as

167. Isocrates, *Panathenaica* 26–32.

168. Seneca, *Moral Epistles* 88 (fully analyzed in Stückelberger 1965 and more briefly in Kidd 1988: 359–65).

169. Cicero, *On the Republic* 1.18.30.

applying only to "some" or "many"), it is clear he would regard as valuable and educated any scientist who was *also* upstanding in all the ways Isocrates describes, and so would anyone agree who shared his opinions.[170]

Hence when Isocrates says the study of science is no longer appropriate for older men, he is again referring only to general education, which for the majority must aim at the mastery of rhetoric and ethical reasoning. So he is not condemning the aims or education of specialists, but insisting that even they become gentlemen.[171] Seneca said very much the same.[172] Even the Roman scientist Galen said that any theoretical science which was "useless" to either moral development, political expedience, or practical life was suitable only for specialists who chose to study such things, and thus was "not necessary" for anyone else to pursue, yet he certainly did not begrudge anyone pursuing them (as he concedes Plato did, whom he was actually defending with these remarks).[173] And this is from one who (as we shall see) was a leading advocate of expanding science education to all who could learn it—and yet who also, like Isocrates, regarded the developing of moral character as the priority of any general education.[174] And it was for that result that Isocrates believed the most general education should emphasize moral philosophy and rhetoric (as the skill of clear-headed and practical reasoning) over the honing of specialized skills in other domains.

Of course, Isocrates' arguments indicate that it was actually commonly the reverse in his time, and therefore a significant element of Athenian society evidently held the pursuit of astronomy and other sciences in even higher regard than he did, although this does not tell us anything about the state of education in the early Roman empire, or even in the Greek world outside Athens in the time of Isocrates. But the evidence suggests that Isocrates' opinions became universal practice: rhetoric and dialectic became, as he wanted, the highest end of all education, while philosophy and science became a mere occasional adjunct to that, or a preparation for it, or the

170. So Poulakos 1997: 100–01.

171. Poulakos 1997: 99.

172. Seneca, *Moral Epistles* 88.36–41 (with other parallels to Isocrates in 88.2 and 88.29–30). See Stückelberger 1965: 46–52.

173. Galen, *On the Doctrines of Hippocrates and Plato* 9.7.9–9.9.14.

174. This was the common view of education, even of the encyclical: Morgan 2011.

educational objective of specialists alone. But as long as such specialists met Isocrates' ideals by achieving a sufficient level of education in the study of literature and rhetoric, thus becoming cultured, well-spoken, and civilized, and as long as their work, whatever it was, ultimately benefitted society and the state rather than serving the aim of mere personal gain, then their work, and the specialized education required for it, was perfectly respectable.[175]

This is confirmed by Roman educational authors. Though there is evidence of some initial Roman hostility to the importation of Greek educational standards in the 2nd century B.C., it did not take long for Rome to fully embrace them.[176] Apart from Varro's evident enthusiasm (and he was among the most respected and revered intellectuals among his fellow Romans), Cicero captured the spirit of Isocrates perfectly (and Cicero, too, came to be highly revered for centuries).[177] Cicero praised Pericles for being the "first orator to be influenced by theoretical research" by studying under "the natural philosopher Anaxagoras," since from this training Pericles "found it easy to transfer that mental discipline from obscure and abstruse problems to the practical business of the forum and the popular assembly."[178] In defending his own educational ideals, Cicero elaborates so well on this point that we should read his remarks in full:

> Let us assume, then, at the beginning what will become clearer hereafter, that philosophy is essential for the education of our ideal orator. Not that philosophy is everything, but that it helps the orator as physical training helps the actor (as it is frequently illuminating to compare great things with small). For no one can discuss great and varied subjects in a copious and eloquent style without philosophy—as for example in Plato's *Phaedrus*, Socrates says that Pericles surpassed other orators because he was a pupil of Anaxagoras the natural philosopher. From him Socrates thinks Pericles learned much that was splendid and sublime, and acquired copiousness and fertility, and—most important to eloquence—knowledge of the kind of speech which arouses each set of feelings. The same may be held true of Demosthenes, from whose *Epistles* one may learn how diligent a pupil he was of Plato.

175. So Isocrates, *Panathenaica* 12 and *Antidosis* 130–36, 174, 285.

176. Bonner 1977: 65–66.

177. On Cicero's educational views see Bonner 1977: 81–89.

178. Cicero, *Brutus* 44.10.

> Surely without philosophical training we cannot distinguish the genus and species of anything, nor define it nor divide it into subordinate parts, nor separate truth from falsehood, nor recognize consequents, distinguish contradictions, or analyze ambiguities. And what am I to say about the knowledge of the nature of things, which supplies a wealth of material to the orator? Again, would you think one could speak or think about life or duty or virtue or morals without thorough training in those subjects, too?[179]

And:

> It is desirable that the orator should not be ignorant of natural philosophy, which will impart grandeur and loftiness, as I said above about Pericles. When he turns from a consideration of the heavens to human affairs, all his words and thoughts will assuredly be loftier and more magnificent. Nor, while he is acquainted with that divine subject, would I have him ignorant of human affairs.[180]

Cicero thus declares natural philosophy important for the orator's education. He also includes dialectic in education as a branch of philosophy, and then satisfies Isocrates' ideals of a moral education by introducing that as moral philosophy, thus employing Isocrates' standards to fold the whole of philosophy into rhetoric.[181] And Cicero's ideals were something of a reality, since he casually remarks how the legend of Pherecydes' prediction of an earthquake (obviously impossible, even to Cicero in the 1st century B.C.) is among the "nonsense" of natural philosophers that is "often heard in schools."[182] And subjects in natural philosophy did come up as practice debates in rhetorical schools of his

179. Cicero, *Orator* 15.3 (citing Plato, *Phaedrus* 269e). Notably Cicero has just described in this paragraph some of the basic skills of Hypothetical-Categorical-Abstract Reasoning (see earlier note).

180. Cicero, *Orator* 119.6 (see also *On the Orator* 1.20, 1.72 and 2.5).

181. Note that the precise division of philosophy into logic, ethics, and physics (meaning the whole study of nature, not just what we mean by physics today) was introduced at least by the time of Aristotle and became routine after Isocrates. Plato is often credited as the first to develop it (see Dillon 1993: 57).

182. Cicero, *On Divination* 2.30.2–11.

day.[183] Ultimately, Cicero's complete list of useful subjects for the orator's education included "rhetoric, ethics, psychology, history, jurisprudence, military and naval science, medicine, and physical sciences such as geography and astronomy."[184] It is clear not everyone agreed with him on the importance of all these subjects.[185] But this only meant his ideals were not universally realized, which we have already observed was the case. A sizable minority of students, however, still held to these high standards. And like Isocrates before him and Quintilian after him, Cicero also argued that statesmen need only know enough not to be ignorant, and should not pursue any science to the degree an expert would.[186] But again this did not represent an attack on experts, only the practical fact that, as Cicero specifically says, the statesman has many other duties, legal and political. The same point is even made by Cicero's scientific contemporaries, the geographer Strabo and the engineer Vitruvius.[187]

The same ideals were echoed by Tacitus (in the second century A.D.), who also argued that true masters of rhetoric will have advanced "in every kind of study" including "geometry, music, grammar" and "every branch of philosophy" including "the subtlety of dialectic, the utility of ethics" and knowing "the causes of the change of things" (*rerum motus causas*)—in other words, natural philosophy. For Tacitus, the orator's excellence comes "from great erudition, a multitude of arts, and a knowledge of all things," because the true orator "is he who can speak on every question" with consummate skill. And so Tacitus praises the great orators of old—with Cicero as his model example—for their "endless labor, daily contemplation, and incessant practice in every branch of study." Hence Tacitus insists the orator must not be deficient in "geometry, music, grammar, or any other liberal art."[188] Tacitus still emphasizes moral philosophy (and moral psychology) above all else, just as everyone in antiquity did.[189] But he also emphasizes the

183. Cf. Cicero, *On Invention* 1.8; *On the Orator* 3.107–10.

184. Gwynn 1926: 101; cf. Cicero, *On the Orator* 1.10.44–1.18.84.

185. Cf. Cicero, *On the Orator* 1.53–57, 2.65–70.

186. Cf. Cicero, *On the Republic* 5.5.14 and *On the Orator* 1.15.65–1.18.84.

187. Strabo, *Geography* 1.1.20–22; Vitruvius, *On Architecture* 1.1.11–18.

188. For all the above: Tacitus, *A Dialogue on Oratory* 30.2–7.

189. Tacitus, *A Dialogue on Oratory* 31.

importance of wide erudition and philosophical eclecticism—expecting the expert to study every major branch and sect of philosophy and take what is useful from each.[190] Of course, Tacitus complains that too many contemporary orators are unlearned and thus fail to meet these high standards.[191] And this no doubt reflects the reality that the advantages to be gained from a successful career as an orator were attracting a mob of 'hacks' to the profession, at least in Tacitus' view, but this would not entail a decline in the number of learned orators, since it would more likely reflect a rise in the number of unlearned orators, who unlike their cultured peers "quite dread the study of wisdom and the advice of experienced men."[192] The same problem of poorly educated pretenders trying to benefit from the popularity of an art is a complaint found in writers for other professional fields, too, from engineering and medicine to even astrology. But this still meant a considerable number were getting the proper education, as Tacitus clearly thought he had, and he could not have been unique in that regard, even if he were in the minority. The significant point is that Tacitus' attitude closely mirrors Isocrates' and yet does not argue for the exclusion of science, mathematics, or natural philosophy from education, but for their inclusion, and though his complaint indicates the failure of ancient society to universalize this ideal, it also reveals that there were still many, like Tacitus himself, who were achieving it.

Cicero wrote in the mid-1st century B.C. and Tacitus in the early-2nd century A.D., while between them, at the end of the 1st century A.D., is the greatest educational writer in antiquity, whom we've met here before: Marcus Fabius Quintilianus, a.k.a. Quintilian. Quintilian elaborates on the ideals and attitudes of Isocrates, and describes the standards Quintilian applied in his own school for over twenty years, which other teachers of comparable excellence must also have enforced. Quintilian argues that "boys ought to be instructed" in "the course of education described by the Greeks as *enkuklios paideia*" before they are "handed over to the teacher of rhetoric."[193] These "other subjects of education must be studied simultaneously with literature" even though they are independent of the study and primary

190. On the rising phenomenon of 'philosophical eclecticism' see chapter seven.

191. Tacitus, *A Dialogue on Oratory* 32.

192. Tacitus, *A Dialogue on Oratory* 32.3.

193. Quintilian, *Education in Oratory* 1.10.1.

aims of oratory.[194] Quintilian is aware of (and responds to) educators who argued these additional studies are unnecessary and useless (especially, his opponents argued, to judges, lawyers, magistrates and politicians), and he is aware of the fact that anyone could "produce a long list of orators who are most effective in the courts but have never sat under a geometrician," or studied music, for instance.[195] He himself knows many educated orators who never pursued these studies, though he despises and derides them.[196] Hence Quintilian admits he has in mind not the typical, but only the best orator when he argues everyone should study these subjects.[197] All of this implies that most students would not be exposed to them, but at least some were, and not so few as to seem bizarre.

Quintilian's advocacy of these studies tells us something about the highest Roman ideals and the attitudes toward science and natural philosophy they inspired. He argues the subjects of the *enkyklios* are useful and always serve to improve an orator's mind, knowledge, and abilities, and if they cannot all be mastered, one should at least study them all as far as one can.[198] Quintilian spends the most time defending the study of music, even though it seems evident he had only a few opponents of that policy in mind (as in practice music was actually one of the most popular educational subjects).[199]

As for geometry, Quintilian argues:

> Elements of this science are of value for the instruction of children: for admittedly it exercises their minds, sharpens their wits and generates quickness of perception. But it is considered that the value of geometry

194. Quintilian, *Education in Oratory* 1.10.2.

195. Quintilian, *Education in Oratory* 1.10.3, 1.10.4, 1.10.8.

196. Quintilian, *Education in Oratory* 1.12.16–18.

197. Quintilian, *Education in Oratory* 1.10.4–8.

198. Quintilian, *Education in Oratory* 1.10.5–8.

199. Quintilian, *Education in Oratory* 1.10.9–33 (he argues music's study and veneration is very ancient, has long been the admirable pursuit of famous and revered men, is closely associated with the divine, has many laudable and useful functions, will improve an orator's ability to control and improve his voice and body language, and is essential to reading and reciting poetry, which was always a central concern of ancient education).

> resides in the process of learning, and not as with other sciences in the knowledge thus acquired. Such is the general opinion.[200]

This means the "general opinion" among Roman-period educators had essentially become that of Isocrates, which continued to laud geometry as a valuable pursuit.[201] Quintilian also defends the study of arithmetic as essential to avoiding embarrassment in certain legal and political contexts, adding that even geometry can be crucial in this respect, too.[202] An orator must also know geometry in order to meet the requirement that he be able to speak on every subject, though he did not need to study it in as much depth as a specialist, who might "dive deep into the minuter details of geometry."[203] Quintilian further argues that the methods established in geometry provide essential foundations for logic and proof that are of fundamental value in all rhetorical contexts, and a strong command of geometry is very often useful in countless other ways, especially by providing a check against many erroneous beliefs—a case for geometry's inclusion in general education that was also made by Galen a century later.[204] Quintilian understood geometry to include the mathematical elements of the study of astronomical theory, which he also says has various uses for an orator.[205] Quintilian thus defended

---

200. Quintilian, *Education in Oratory* 1.10.34.

201. Quintilian also includes training under an actor among the ancillary studies (*Education in Oratory* 1.11.1–14), for perfecting such things as presence, diction, and delivery, but it is unlikely he imagined this to be a part of the *enkyklios*. Likewise, he gives a nod to gymnastics (ibid. 1.11.15–19), only as far as studying under a 'movement coach' to perfect bodily grace and other uses of body language, since he disapproved of further attention to gymnastics, though implying at the same time that gymnastics was more commonly a standard element of Greek education (see earlier note on educational athletics).

202. Quintilian, *Education in Oratory* 1.10.35–36. Cicero makes the same point about ignorance of the liberal arts causing undesirable embarrassment in *On the Orator* 1.16.72–73.

203. Quintilian, *Education in Oratory* 1.10.49 and 1.12.14.

204. Quintilian, *Education in Oratory* 1.10.37–49. See discussion in Cuomo 2000: 47–48 (Quintilian) and Cuomo 2001: 187–88 (Galen).

205. Quintilian, *Education in Oratory* 1.10.46–48; cf. also 1.4.4. The division of astronomy into its mathematical and physical aspects is a phenomenon I'll discuss in *The Scientist in the Early Roman Empire*, but may have been influenced by the

the entire *enkyklios* as the best education of his day and the standard everyone should aim for as far as they were able, and at least some achieved. Like Tacitus and Cicero, Quintilian went beyond the *enkyklios* to advocate the whole of natural philosophy as a worthy subject of study, but we shall discuss his remarks on that point later.

Apart from defending their practical uses, Quintilian echoes Isocrates when he argues that subjects like geometry and music theory are better distractions for juveniles than the trouble they could get into as delinquents.[206] He also argues that these topics served as a useful way to break the monotony of literary studies, thus benefitting education as a whole.[207] And just as Aristotle frowned upon pursuing them for mere gain, Quintilian praised the "stable and lasting rewards" that only "knowledge and contemplation" of such subjects can bring—rewards, he believes, that will win a man's heart away from vice and turn him to nobler occupations and entertainments.[208] This was not an uncommon view. The Jewish philosopher Philo, writing early in the 1st century A.D., composed a highly mystical and allegorical treatise on the *enkyklios*, arguing throughout that all seven subjects ought to be pursued because they are necessary for the acquisition of virtue. As Philo insists, "we cannot attain virtue until we attain its handmaiden," the *enkyklios*, which he calls "the road to virtue."[209] Likewise, Cicero argued that

---

fact that the mathematics could be taught by a geometrician aiming foremost to teach abstract principles while the astrophysical part could be taught by an astronomer aiming foremost to teach specific facts and practices. That both aspects of astronomy were nevertheless a common part of education is further implied by Cicero in *On the Orator* 1.35, 1.128, 1.149, 1.158, 1.187, 2.28. The same implication follows from Seneca's remarks in *Moral Epistles* 108.1, 114.10–19, 115.1 (and most of epistle 88).

206. Quintilian, *Education in Oratory* 1.12.18.

207. Quintilian, *Education in Oratory* 1.12.4–5, 1.12.13–14.

208. Quintilian, *Education in Oratory* 1.12.16–18; compare Aristotle, *Politics* 8.2.1337b.

209. Philo of Alexandria, *On Mating with the Preliminary Studies* 3.9–11, 4.14–18, 14.74–79 (throughout he explicitly names only six of the standard seven, omitting arithmetic, but he adds arithmetic to geometry and harmonics in *On the Special Laws* 2.32.200 and *On the Creation of the World* 37.107); and Philo, *On Agriculture* 3.14–4.20 (which also adds natural, moral, and dialectical philosophy in making the same point; cf. also Philo, *On the Change of Names* 10.70–76). In *Mating* Philo uses

teaching science to the wider public would "banish fear and false religion from confused men" and thus would be a positive good, which many agreed would have practical benefits to the state.[210] For example, using science lessons to better manage soldiers, a technique found not only in Cicero, but explicitly described in Roman manuals on military strategy.[211]

Modern scholars of ancient education think geometry was 'mostly' of interest to specialists, and would be seen as a kind of special vocational track in secondary school, pursued if one planned to seek an apprenticeship or education in a field that required it (like philosophy, astrology, surveying, engineering), although the remarks we've seen from ancient authors entail many among the wealthier classes would also have pursued it simply out of personal interest or the high expectations of their parents or patrons. Philo, for example, reports that he studied the *enkyklios* in his youth because of his desire to study philosophy, but he knows of many people who do not continue on to philosophy but stop at completing the *enkyklios*, or continue further in one of its subjects.[212] But this entails such an exceptional education had to have been widely available, and just not *universally* pursued.[213] Another example is Nicolaus of Damascus, the famous Aristotelian (and court historian to Herod the Great), who wrote in his autobiography that since he was a boy he had avidly studied music and "the theory of mathematics" (by which he most probably meant the remainder of the quadrivium:

---

the 'handmaiden' theme to produce allegorical interpretations of various biblical passages and stories, especially Abraham's 'conjugal' relations with Sarah and Hagar. For more on Philo's views see chapter nine (I will further explore his views in *Scientist*, where I shall also discuss how Philo's 'handmaiden' idea was later adapted by Christians to subordinate the whole of philosophy to the gospel, which is also touched on briefly here in chapter nine).

210. Cicero, *On the Republic* 1.15.24–1.16.25. Examples of others lauding its value to the state: Valerius Maximus, *Memorable Deeds and Sayings* 11.1; Plutarch, *On Superstition* 8 (= *Moralia* 169a-b).

211. Frontinus, *Stratagems* 1.12, who entitles an entire chapter "On Undoing the Fear Which Soldiers Derive from Adverse Omens," where seven examples involve manipulating the superstitions of soldiers (§ 1, 2, 4, 5, 6, 7, 12) but three examples are of using science lessons to the same end (§ 8, 9, 10).

212. Philo of Alexandria, *On Mating with the Preliminary Studies* 14.74–79. See Sandnes 2009: 68–78.

213. Clarke 1971: 47–49.

arithmetic, geometry, and astronomy), as well as "grammar," "rhetoric," and "all of philosophy" (which would thus include dialectic, completing the trivium).[214] Like Philo, he mentions how one could study each of these subjects to differing degrees as one desires, and that all were available to study. So clearly many among the elite were doing this.

Although this still meant that for *most* students mathematical education never went beyond basic arithmetic and included "minimal concern for developing critical thinking and creative ability."[215] And as for geometry, so for science education generally (although in all honesty, modern American education fares little better by this measure, and it may be that no general education system ever has). However, it is possible this disparity has been exaggerated, insofar as these conclusions rely mostly on evidence from extant school exercises preserved on papyrus and potsherds. For we cannot expect to have comparable physical evidence of school work in geometry, since it was taught with a sandbox and by a specialist, and teachers and students of geometry would not have consumed costly papyrus, or ink on ostraca, when they already had a suitable medium for their work.[216] The physical evidence therefore must underrepresent how often geometry was actually studied. Though by its nature it was still more expensive, less necessary, and not universally advocated or required, and thus a majority of grammar students will not have studied geometry, nevertheless the minority who did was probably larger than the physical evidence has been taken to suggest. In a similar fashion (as already noted in the previous chapter), "pupils who wanted to learn rhetoric seem typically to have moved to large towns and

214. *FGrH* (*Die Fragmente der griechischen Historiker*) 90.F132 (= Suda, s.v. "Nikolaos" [*nu* 393]).

215. Cribiore 2001: 180–84.

216. On the use of sandboxes in geometry: Clarke 1971: 51–52; Bonner 1977: 77; cf., e.g., Seneca, *Moral Epistles* 88.39. Other educational aids existed. Wax tablets, often in codices of multiple leaves, were common equipment for ancient students. And Lucian, *Nigrinus* 2, describes the classroom of a Platonist teacher, who had among his teaching aids "a [wax] tablet filled with figures in geometry and a reed sphere, apparently made to represent the universe." On the use of similar spheres in education see Clarke 1971: 52. Wealthier teachers could have elaborate armillary spheres for use in their lectures (e.g. Theon of Smyrna, *Aspects of Mathematics Useful for Reading Plato* 3.16.146; I'll discuss the use and manufacture of such advanced instruments in *The Scientist in the Early Roman Empire*).

cities to do so, and these are places from which texts composed by pupils do not generally survive," a fact that would also skew the evidence regarding what actual and prospective students of rhetoric actually studied, or even hide disparities in focus between urban and rural students generally.[217]

These problems do not change the overall conclusion that pursuit of the full *enkyklios* was not the majority experience in ancient education. As Quintilian and Gellius both observed, too many students for their liking were entering higher education without it.[218] Yet Plutarch observed enough students completing the full *enkyklios* to confirm it was not exceptionally rare. He even expected students to complete some philosophical training at the secondary level simply to prepare them for exposure to philosophical subjects in higher education, and he clearly knew many students who had done this.[219] This probably meant completion of the *enkyklios*. For he knew many students who not only entered higher education prepared with a prior training in logic and mathematics, but who would annoy professors by interrupting their lectures with advanced questions or complex mathematical or logical problems, simply to show off their expertise in these subjects.[220] These rambunctious teens surely represented only a fraction of the otherwise well-behaved students who had completed the *enkyklios*, and yet these braggarts were clearly common enough to become a regular problem for professors, hence an example Plutarch could use for what a well-behaved student should not do.

The *enkyklios* and its science content were thus pursued by a great many students in antiquity, even if they were only a minority among the elite. By comparison, how many youths studied the *enkyklios* and its basic science content in the Middle Ages? I suspect it is not likely even to be comparable, much less greater. But I must leave that for others to determine.

---

217. Morgan 1998: 7.

218. Aulus Gellius *Attic Nights* 1.9.6–7; Quintilian, *Education in Oratory* 1.12.16–18 (with 1.10.3–8). Galen, too, says "most people pretending to some education" have not studied geometry: *On the Uses of the Parts* 10.12 (= May 1968: 490, 492). Nevertheless, many nonscientists did study mathematics and mathematical sciences (cf. Netz 2002: 210–13 on Polybius as an example).

219. Plutarch, *On Listening to Lectures* 2 (= *Moralia* 37f).

220. Plutarch, *On Listening to Lectures* 10 (= *Moralia* 43a-b).

6. Higher Education

Whether completing the *enkyklios* or not, the next stage of education was to attend the lectures of a professor. As this began in one's mid to late teens, ancient "higher education" could be thought of as overlapping the latter part of what we now call "high school" in America. But many could begin such studies at a later age, depending on their circumstances. Limits on who could attend were created by the need for books and advanced reading and writing skills and materials, as well as the high fees, and the rarity of both professors and opportunities to become their student (in both numerical and geographical terms). These difficulties conspired to lock out most of the ancient population by far. And yet for most students, who did not complete the *enkyklios*, most if not all exposure to science (all serious facts, theories, methods, and concepts in the study of nature) took place only in higher education. Moreover, the predominant form of higher education consisted of the study of rhetoric—which included a full range of skills in public speaking and argument, but did not involve very much attention to science. The main purpose of rhetorical schools was a preparation for service as an advocate in government and courts of law.[221] This is repeatedly assumed by Eumenius in his appeal for the rebuilding of a school at Autun,

221. As most forcefully argued in Parks 1945, who also advances a useful running counter-argument against more negative assessments of the 'Second Sophistic' (on which see note in chapter four). For a full survey of the aims and content of an ancient education in rhetoric see Gunderson 2009, Morgan 2007, and Wouters 2007 (supplemented by Panella 2011–2012, Walker 2011, and Brodie 2004: 2–79). And on law as a profession (for which one certainly needed an education) see Kleijwegt 1991: 165–86.

for example.[222] Thus science was not a focus, any more than it tends to be in modern law schools.

Even what limited scientific content a student of rhetoric would be exposed to would mostly enter the curriculum only in later years, and yet not all students at this level completed a full course of study. Many finished just enough schooling to get ahead in society, which generally meant they often dropped out before getting any more significant exposure to scientific knowledge.[223] For example, Seneca says his friend Lucilius commendably studied philosophy and the liberal arts despite his poverty and the temptation only to study rhetoric just enough to make money.[224] Similarly, in his *Teacher of Rhetoric* Lucian parodies teachers who offered an accelerated (and thus absurdly abbreviated) rhetorical training for the same purpose.[225] The preliminary years of rhetorical training also remained focused on poetry, only gradually moving to prose, which became a significant element of the curriculum only in the final years, and even then, science content would still mostly have been in the form of incidental comments, digressions, and analogies, and incidental zoological and geographical detail, since history and oratory were the primary genres studied, rather than actual scientific, technical, or philosophical treatises. In Greek, Demosthenes was always studied intensely, but also Isocrates, then Hyperides and Aeschines, sometimes Lysias and Lycurgus, and only afterward were historians considered, the most popular in use being Herodotus, and at more advanced levels of study, Theopompus, Xenophon, Philistus, Ephorus, and Thucyidides.[226] The typical Latin curriculum did not differ substantially in the kind of books employed.[227] The science content of such works was

222. Eumenius, *For the Restoration of the Schools* (= *Latin Panegyrics* 9), regarding the school at Autun, Gaul, c. 298 A.D. (after the devastating events of the 3rd century). See Cribiore 2007 for similar evidence in Lucian and Libanius. On the evidence of Eumenius: La Bua 2010.

223. See Cribiore 2001: 220–44 and Marrou 1981.

224. Seneca, *Natural Questions* 4a.pr.14.

225. See Cribiore 2007.

226. Cribiore 2001: 225–38, 231–44.

227. See Quintilian, *Education in Oratory* 1.8.5–12, and 10.1, where the emphasis is on poets, orators, and historians, in that order—though he does include some philosophy, that would not have been common.

minimal Herodotus, for example, was one of the most popular historical texts studied at this level,[228] and it included some geography and natural history and a few doses of the fantastical, but almost no science.[229] Thus, even at best an education in rhetoric did not contain a great amount of scientific content.

Hence all significant science education took place elsewhere, either in schools of philosophy, which only a minority of students at this level attended, or under the tutelage of actual scientists, which was available to fewer still. Those who decided to pursue practical or academic careers in philosophy, medicine, astrology, or engineering, rather than focusing all their energies on a generic rhetorical education, would come away with a far more impressive science education than anyone else in the Roman empire, though in varying degrees. As noted in chapter one, this differs considerably from today, since the Scientific Revolution has transformed our educational values. Now science is a major pursuit in colleges and universities, and has become a fundamental element of all school curricula even at the elementary level and especially at the secondary level and beyond. Thus scientific knowledge, and what was in antiquity advanced mathematics (geometry, algebra, and trigonometry), is today an educational mainstay, exactly the opposite of antiquity, where literacy and oratory and the skills related to reading, composition and argument represented the primary educational values, and most preparation for higher education consisted of the mastery of fundamentals related to those pursuits, leaving science as an ancillary or specialized interest even at the highest levels. But I suspect very much the same could be said of the Middle Ages.

However, Marrou was incorrect to conclude that "science played only a small part in Hellenistic education" and "was taken up by an even smaller minority of specialists than philosophy."[230] Because the *enkyklios paideia* had considerable science content (as demonstrated in the previous chapter) and would have been pursued by more people than even went on to higher education at all, much less to study philosophy. And as we

228. Cribiore 2001: 144.

229. Sergueenkova 2009.

230. Marrou 1956:254 (= Marrou 1964:372,"les études scientifiques n'occupaient que peu de place dans l'éducation hellénistique: plus encore que la philosophie, elles n'intéressaient qu'une infime minorité de spécialistes").

shall see in chapter seven, all the major philosophy schools taught some measure of science, too. But Ben-David is right that "science was virtually absent from the curricula of the rhetorical schools which were the most widespread educational institutions" of the time.[231] And Cribiore concurs. "Science" in the sense of "higher mathematics, geometry, or astronomy," and "philosophy" were taught by "specialized instructors" and were "not part of standard education" but instead "attracted a very restricted student population."[232] This is confirmed by Apuleius, who says most do not go beyond rhetoric nor even finish the *enkyklios*, and Diodorus, who says most who attend schools of rhetoric eventually give up their studies to earn a living instead, while only a few go on to study philosophy (or anything more advanced).[233]

Although again the conclusion must not be taken too far, since in absolute numbers this could still mean quite a lot of students, e.g. even if only 1 in 5 students at the secondary level completed the *enkyklios*, that would still amount to tens of thousands in any generation. Beyond the *enkyklios* the numbers would be thinner. But natural philosophy did not become a significant element of any *standard* higher education in the West until the 12th century at the earliest, and even then it only gradually developed a broader impact.[234] This must have had some impact on the future of science, but the Scientific Revolution still did not arrive for another three hundred years, so whatever that impact was, it clearly was not sufficient to cause a revolution, unless by a gradual snowball-effect it actually required centuries to have its overall consequence—by, for example, slowly growing the number of persons both interested and competent in the sciences until their number reached a critical mass that tipped the scales of social values.[235] However, it is not yet clear whether the number of persons exposed to a university

231. Ben-David 1984: 42.

232. Cribiore 2001: 3.

233. Apuleius, *Florida* 20 Diodorus Siculus, *Historical Library* 2.29.5–6.

234. Beaujouan 1963 (with relevant discussion in subsequent scholarship on medieval universities, previous note).

235. Essentially argued in Rihll 2002: 12–15, Collins 1998: 523–69, Crombie 1963: 9, and Edelstein 1952: 598–99 and 1963: 30–32 (although see my estimate of numbers in chapter three; I'm unaware of any comparable estimate attempted for the Middle Ages or the Renaissance). I will discuss this theory further in *Scientist*.

education in the High Middle Ages or even the early Renaissance was substantially greater than the number of persons exposed to an education in philosophy or the sciences in antiquity, and this question cannot be resolved here. Mere attendance alone would be an insufficient measure anyway, since the actual science content of medieval university curricula was very limited and circumscribed. In fact, if we are only talking about a *standard* university education, then medieval science content did not go much beyond what the ancients were already teaching in the *enkyklios*. Thus medieval higher education should not be compared with ancient higher education alone, but ancient higher *and encyclical* education together.

There is also no evidence of a declining importance of the sciences in ancient education, either before or during the early Roman Empire, as has been claimed by Marrou and Grant.[236] Neither of them present any convincing evidence that science was prominent in education at *any* time in antiquity. By advancing only two or three unrepresentative and idiosyncratic examples wholly insufficient to establish a generalization, they latch onto the unjustified assumption that science was somehow more pervasive in Classical or early Hellenistic education, and thus when the evidence becomes more substantial (with the expansion of Hellenistic education throughout what would become the Roman empire), they perceive an illusion of decline, when in fact the education that became universalized across Europe and the Mediterranean was essentially identical to the education most widely pursued from the start. Just as there were exceptional cases before that expansion, there continued to be exceptional cases after it, allowing no

---

236. Marrou 1964: 274–77 (= Marrou 1956: 182–84) and Grant 1952: 78. Likewise, the oft-repeated myth that Romans only sought practical applications for mathematics and abhorred mathematical theory, while the Greeks did the opposite (e.g. Gwynn 1926: 18–92), is attacked by Marrou (1965: 410 = 1956: 282), criticized by LeHoux (2012: 2–8), and completely refuted by Cuomo (2001: 192–211), who also provides an apt analysis of how and why Greek educational ideas were assimilated into Roman culture. Eyre's claim that "the Romans were *not* interested in knowledge for its own sake" and conducted "no research" and developed "no technical or commercial education" (Eyre 1963), is not only false but ridiculous (in fact most of his claims are long on assertion but short on evidence). Native Greeks did maintain a linguistic advantage in the sciences (as I explained in chapter three), but "Romans" were just as enthusiastic about theoretical knowledge and inquiry as the Greeks, and Greeks were just as interested in practical applications as the Romans.

basis for generalizing any imagined decline in attention to mathematics or science. Such attention was *always* minimal and exceptional. Similarly, the *quality* of even the limited science content in ancient education, at all levels and in all historical periods, was never high relative to what specialists knew and taught even then, much less now.[237]

True, we hear a complaint about Roman teachers so occupied with competing for students that they downplayed the teaching of boring or profound subjects and instead emphasized topics, examples, and analogies that were either entertaining or frivolous, yet there is no good evidence that anything had ever been significantly different in this regard, nor any reason to believe anything would have been.[238] The social and economic demands on teachers had never changed, nor had the interests of the privileged class. A substantial increase in the availability of education compared to the Classical period, which must have taken place under the increasing expansion and economic benefits of the Hellenistic and Roman periods, would have the effect of increasing the market for cut-rate teachers and half-educated pretenders, even while the availability of a quality education remained as small as it had ever been. Though the proportion of the badly educated to the superbly educated may have remained the same, the overall increase in numbers would make the rise of badly educated students far more visible and alarming, creating the subjective impression of a decline. However, when we survey what objective evidence there is of what the literate were actually teaching, studying, or knew, we look in vain for any supposed heyday of widespread science education in the ancient world. Thus, while Bonner hypothesizes that "the more urgently a teacher felt the need to prepare his boys to demonstrate their ability in declamation, the less ready he was to recommend the study of subjects of less immediate relevance, such as arithmetic, geometry, musical theory and astronomy," this had *always* been the case, even at Athens in the days of Protagoras or Isocrates.[239] It made no greater difference under the Romans.

---

237. For this problem in the Roman Republic: Rawson 1985: 156–69, 287–88.

238. On the complaint see Tacitus, *A Dialogue on Oratory* 29 (discussed in chapter five) and Galen, *On My Own Books* Kühn 19.9. The latter is certainly hyperbolic, since Galen asserts that Greeks "always" used to be taught letters and grammar, which was certainly never the case.

239. Bonner 1977: 102–03.

All that aside, schools of rhetoric would usually be the first time a student would learn any significant skills of argument and reasoning, including a basic arsenal of logical and rhetorical forms with which to criticize or refute an opponent or to argue for and defend a position (as well as skills pertaining to oral and literary creativity). The ultimate aim of such education was "an almost unfailing precision in analyzing the pros and cons of cases," mainly as would arise in legal or political debate.[240] Students who completed grammar school and got no further will have received almost no such experience. But if a student continued, then a full course of study in rhetoric typically took five to seven years, and most commonly began in a student's mid-teens.[241] Many professors required students to pass an oral "entrance exam" of sorts to demonstrate their adequate preparation.[242] But standards varied, and there were many lousy or ignorant teachers, which would not have helped in the dissemination of sound natural philosophy, or the reliable transmission of a natural philosopher's epistemic values, even as occasions arose in oratorical commentary and discussion.[243] But the best educators still insisted on as much study of natural philosophy as was reasonable and useful. Quintilian, for example, declared that "no one will achieve sufficient skill even in speaking, unless he makes a thorough study of all the workings of nature and forms his character on the precepts of philosophy and the dictates of reason," referring to all three branches of philosophy as an interrelated whole.[244]

In fact, Quintilian argues, "natural philosophy is far richer than the other branches of philosophy" in providing lofty subjects to speak on; it is essential to many important political and legal debates about such things as the efficacy of divination and oracles and other matters of religion; and moral philosophy requires a knowledge of human nature and the function

240. Cribiore 2001: 222–25.

241. Cribiore 2001: 56.

242. Cribiore 2001: 224.

243. Cribiore 2001: 57–58.

244. Quintilian, *Education in Oratory* 12.2.4, i.e. "philosophy falls into three divisions, physics [*naturalem partem*], ethics [*moralem* (*partem*)], and dialectic [*rationalem* (*partem*)]" and all are important to the orator, ibid. 12.2.10, including "inquiry into the causes of natural phenomena," ibid. 1.pr.16. On these divisions of philosophy see note in chapter five.

and purpose of the universe, which only natural philosophy can provide. Besides all that, Quintilian asks, "how can we conceive of any real eloquence at all proceeding from a man who is ignorant of all that is best in the world?" And here he repeats the examples Cicero had offered, as we saw above, of Pericles and Demosthenes, and Quintilian now adds Cicero himself, "who often proclaimed the fact that he owed less to the schools of rhetoric than to the walks of the Academy" and would "never have developed such amazing fertility of talent if he had bounded his genius by the limits of the forum and not by the frontiers of nature herself."[245] An example of an orator's need for natural philosophy is offered by Aelius Theon of Alexandria (writing a generation or two before Quintilian), who proposes an abortion case in which the physiology of a fetus becomes a point at issue.[246] There was thus an evident value for natural philosophy in the best schools of rhetoric, and in fact the best orators were expected to study it at least enough to speak intelligently on the subject.[247] And competition for favor, prestige and excellence would drive many to follow their lead.

Thus, in ancient higher education science was still a major educational value in the sense of being part of the "ideal" education and an actual pursuit of the more ambitious or talented students. It was *not* a major educational value only in the sense of not being a routine or universal component of the ancient equivalent of 'going to college'. Nevertheless, as noted earlier, any comparison with medieval higher education will not be so straightforward, as what medieval universities taught to grant a 'bachelor's degree' was essentially the *enkyklios paideia* (discussed in chapter five), which in antiquity was studied in parallel to grammar school, and thus prior to what we mean here by higher education. Antiquity thus allowed two different scholastic tracks: a fast track from grammar school straight to oratory, argument, and law, and a more rounded track that was directly equivalent to what one learned as an undergraduate at a medieval university. Assessing the relative numbers of students in either track as compared with

245. Quintilian, *Education in Oratory* 12.2.20–23.

246. Aelius Theon of Alexandria, *Preliminary Exercises* 2.69 (see Bonner 1977: 83). For other examples of medical science in rhetorical exercises see Gibson 2013 and Ferngren 1982 (esp. pp. 280–81). And see my discussion of juries in chapter four.

247. Quintilian, *Education in Oratory* 1.pr.16–18.

college students in the Middle Ages has not been convincingly done and thus no argument can proceed from an assumption of any difference. There may easily have been many more well-rounded orators in antiquity than in even the High Middle Ages, leaving no fewer students exposed to the sciences.

7. ADVANCED EDUCATION

The ancient equivalent of 'going to college' meant either attending lectures in rhetoric and logic or attending lectures in philosophy—or both, either in that order, or simultaneously.[248] But philosophy was always considered a higher calling that fewer studied. Hence I shall consider it here as an equivalent of 'graduate school', alongside or in addition to the specialized study of an actual science (like medicine, astronomy, or engineering). I will discuss schooling in philosophy first (as this was far more commonly pursued), and then the education of scientists specifically.

Although the best orators were expected to know enough science and natural philosophy not to look like a fool, in philosophy extensive exposure to such knowledge was more routine. The leading schools of philosophy differed as to the extent and kind of natural philosophy that would be taught or expected of their students, in ways too complex to enumerate in detail. But in brief, the leading schools of philosophy were, in order of apparent popularity (at least among the elite) in the Roman period (according to what seems to have been the relative number of adherents and advocates, and the relative frequency of praise or admiration found in the sources): the Stoics, the Platonists (and Pythagoreans), the Aristotelians (or 'Peripatetics'), the Epicureans, the Skeptics (including both the Pyrrhonists and the Academics), and the Cynics. There was a smattering of minor sects in addition to these, but evidence for any of them in the Roman period is scanty or problematic. More importantly, eclecticism had become very popular in the Roman era, and thus most students of philosophy never aligned with any one school

248. Morgan 1998: 193; Bonner 1977: 82–83.

but studied under several, and many would assemble their own unique philosophy from elements of them all.[249] This is obscured by paying too much attention to those few who declared themselves for a single sect, and then mistaking that minority behavior for the standard. In fact among the majority of philosophy students, eclecticism, not sectarianism, seems to have been the norm. And yet even sectarians studied several sects before settling on any. And they all came into constant public debate with each other. Thus, the attitudes of any one school were not very important, since the average student would be exposed to several schools, at least one of which would involve or require some instruction in science and natural philosophy. The exceptions would be those remaining students who dogmatically attached themselves to only one school right from the start, or who favored certain schools over others in their studies and pursuits. But even then they would be exposed to more natural philosophy than any mere orator, and the more so as they debated with adherents of other schools. With all that in mind, we can summarize the status of science and natural philosophy in each sect.

In general, Roman-period philosophy schools taught or required the study of science or natural philosophy in varying degrees and for various reasons.[250] The Stoics required the study of natural philosophy by every student, taught certain fundamental epistemic values that were highly favorable to science, and promoted a deeper pursuit of science by the ideal 'wise man', though still elevating moral philosophy above all.[251] The

249. See Dillon & Long 1988 (briefed in *OCD* 483, s.v. "eclecticism"); Gottshalk 1987: 1164–71. The best ancient example is the Roman doctor Galen, *On the Affections and Errors of the Soul* 1.8 (= Kühn 5.42–43; also 2.6–2 = Kühn 5.96–103); on which see Hankinson 1992. The Roman astronomer Ptolemy was likewise an eclectic (Huby & Neal 1989; Long 1988), as was the Roman engineer Hero (Tybjerg 2005). Other good examples of this principle being expressed by Roman intellectuals include Seneca, *Moral Epistles* 33 and Celsus, *On Medicine* pr.45–47. This trend became so popular among Romans that an actual 'Eclectic' sect was formed in the reign Augustus (at the end of the 1st century B.C.): see *OCD* 1199 (s.v. "Potamon (2)"), *EANS* 693 (s.v. "Potamon of Alexandria"), and Diogenes Laertius, *Lives and Opinions of Eminent Philosophers* 1.21.

250. On the comparable attitudes of all these sects toward the *enkyklios*: Rawson 1985: 182. On Hellenistic developments in the demarcation and popularity of philosophical sects leading into the Roman era: *OCD* 657–58 (s.v. "Hellenistic philosophy").

251. The standard Stoic curriculum in natural philosophy is summarized in

Platonists also required the study of natural philosophy by every student, but mainly the mathematical sciences, and primarily in their mathematical more than empirical aspects, often emphasizing contemplation over what we would recognize as genuine 'science'. Nevertheless, the student of a Platonist would gain considerable exposure to mathematics and mathematical explanations of natural phenomena.[252] The Aristotelians, on the other hand, comprised the pro-scientific sect *par excellence*, requiring the study of natural philosophy of all its students, to a greater and broader extent than any other sect, and coming closest to teaching the most successful scientific methods and values.[253] The Epicureans also required their students to study natural philosophy, since it was essential to their moral philosophy (a view shared by the Stoics), but the nature and aims of the natural philosophy they taught or insisted upon were not always aligned with the actual sciences of antiquity, especially since the Epicureans did not as much favor the learning or use of mathematics, did not insist upon detailed study or even

---

Diogenes Laertius, *Lives and Opinions of Eminent Philosophers* 7.38–160 and partly reflected in Seneca's *Natural Questions*. See also Brunschwig 2007, Sellars 2006, Evans & Berggren 2006:23–27, Inwood 2003, Morford 2002:161–239, Stückelberger 1988: 35–38, and Edelstein 1967: 137–38, 145, 167–78; along with *OCD* 1403–04 (s.v. "Stoicism").

252. Illustrated by the textbooks on the quadrivium by Nicomachus (see note in chapter five) and remarks in Roman-era introductions to Platonism, like Alcinous, *Epitome of Platonic Doctrine*. See also Kalligas 2004, Joost-Gaugier 2006, Remes 2008, Chiaradonna & Trabattoni 2009, Gerson 2013, and *OCD* 1007–08, 1155–58, 1245–46 (s.v. "Neoplatonism" and "Neopythagoreanism"; "Plato (1)" and "Platonism, Middle"; and "Pythagoras (1), Pythagoreanism"). However, on the probable obsolescence of the "Middle" and "Neo" terminology: Catana 2013. On the role and influence of Pythagorean thought in Roman-era Platonism: Joost-Gaugier 2006.

253. This is readily apparent in the pervasive body of Aristotle's works that remained in circulation (and not only his own, but those of his pupils and successors), many of which survive to this day. See also Boylan 1983, Gottschalk 1987, J. Barnes 1995: 105–67, Falcon 2013, *OCD* 1108 (s.v. "Peripatetic school"), and *EANS* 142–45 (s.v. "Aristotle"), cf. also *EANS* 145–53. Note that very little of what ancient Aristotelians actually taught (both in terms of scientific content *and* scientific methods) was reflected in medieval university curricula (where surviving school texts were startlingly few, obsolete even by ancient Roman standards, and interpreted very oddly). I will discuss what *ancient* Aristotelianism was really like in *The Scientist in the Early Roman Empire*.

literacy, and were more content with metaphysically satisfying answers to scientific problems than other sects were (as opposed to actually testing their theories).[254] Nevertheless, the student of an Epicurean could expect to acquire considerable, if superficial, exposure to the scientific knowledge of their day and to many, though not all, of the epistemic values essential to science.

The Skeptics (a broad category of those who came from a variety of sectarian origins, from Pyrrhonists to Academics) ostensibly rejected all facts about nature as unknowable and thus did not ostensibly value the study of science or mathematics, except to critique and refute them, which ironically entailed studying them extensively, a pursuit that in some cases actually tempered their skepticism *in favor* of some scientific research and results. And in practice most Skeptics accepted the pursuit of 'likely truths', which also led them to various scientific interests and the acceptance of many scientific conclusions. Moreover, they taught students to be very demanding of evidence and argument in the face of any claim about the natural world, which could be of considerable aid to science.[255] In contrast, the Cynics completely rejected all science and natural philosophy as vain and useless, perhaps even dangerous or impious, promoting instead a countercultural abandonment of almost all the aims and trappings of civilization.[256]

Stoicism came to be most respected among the ruling elite in the early Roman empire, and then popular focus shifted over the 3rd and 4th

---

254. See O'Keefe 2010, Warren 2009, Di Muzio 2007, Asmis 2004 (with Too 2001: 209–39), Ferguson & Hershbell 1990, and Edelstein 1967: 135–37, 160–65 and 1952: 594–96, as well as *OCD* 513–14 (s.v. "Epicurus") and *EANS* 287–89 (s.v. "Epicurus of Samos"). Exactly how much science Epicureans taught is a vexed question, but a general idea of its content in the Roman period is probably reflected in Lucretius, *On the Nature of Things*, though there were always exceptional Epicureans who studied more than the school required (including advanced mathematics, a fact I shall examine in *Scientist*).

255. All this is evident from the extant collected writings of Sextus Empiricus and the philosophical essays of Cicero, although Skeptics disagreed with each other on what to teach regarding the sciences: Edelstein 1967: 165–67. I will treat the relationship between ancient Skeptics and science in more detail in *Scientist*. For a general study: Thorsrud 2009.

256. See Desmond 2008 with *OCD* 402–103 (s.v. "Cynics") and Edelstein 1967: 58–63.

centuries to Platonism, which in the end completely eclipsed every other school of philosophy except the Aristotelian, which it still nearly drowned. The other schools in the early Roman empire enjoyed wide respect and some measure of popularity, except the Cynics, who were always a disdained fringe movement with little or no social standing or influence,[257] and the Skeptics, who were regarded with more ambivalence but still enjoyed only limited popularity. Thus, while most students who undertook the study of philosophy would gain exposure to several sects, especially the most popular ones, most of those who settled on one sect still embraced one of the most popular, and since all such sects taught or required the study of science and natural philosophy, this means most students of philosophy—most by far—received a fairly decent science education, in many cases the best that any layperson could then expect. This of course still meant 'science' in the ancient sense, which was often filled with as much error and poppycock as genuine facts and sound theories, but in terms of social values, schools of philosophy by and large embraced and promoted a basic science education and many of the corresponding scientific values.

How a student of philosophy learned science varied considerably. Aulus Gellius reports that philosophical education was highly variable in this respect.[258] How much natural philosophy, much less of any scientific kind, that a philosophy student learned depended on which subjects and, as we have seen, which schools of thought the student pursued most fervently, and what the interests and demands of his particular teachers were. Plutarch says there were specialists in natural philosophy who should not be expected to be experts in dialectics or mathematics, and vice versa, and that one would attend the lectures of many different experts like these, and accordingly he advises students to restrict their questions to what the speaker knows.[259] Overall, studying philosophy meant to some extent studying the sciences sufficiently for a reasonable lay understanding. The more so as the popularity of philosophical debate demanded it. It could be socially, economically,

257. Notably Socrates in his attitude toward the sciences was closer to the Cynics than any other school that came after him (e.g. Xenophon, *Memorabilia* 4.7; cf. Prince 2006 and McKirahan 1994), but like Cynicism as a whole, his attitude was almost universally rejected by the elite in antiquity.

258. Aulus Gellius, *Attic Nights* 1.9.6–8.

259. Plutarch, *On Listening to Lectures* 11 (= *Moralia* 43c).

and psychologically disastrous to be exposed as an ignoramus in a public confrontation, necessitating a sufficient familiarity with the scientific facts that could be summoned in support of any argument. A few Christians got wise to this very same problem (at least Clement and Augustine, as I'll discuss in chapter nine), but their advice as to solving it was generally ignored. And of course not every pagan philosopher heeded such advice, either. But most must have, and most extant texts by and about Roman era philosophers confirm this.

For this reason science and logic became inexorably linked: it was futile to debate and study logic if one's premises were easily challenged by scientific experts and facts. This was similar to the way ethics and science became inseparable, because moral philosophy required establishing the facts of human nature and the way the universe worked, but also because it was widely believed scientific knowledge lifted the spirit above base concerns and was thus essential to developing a moral mind.[260] In the same fashion, logic could not be pursued in ignorance of science, while science became a battleground for debates in logic, due to the sciences' many peculiar and pressing problems in epistemology and persuasion. Roman scientists were thus fully engaged in this rage for logic, Galen writing treatises in logic, Sextus surveying the whole field, Hero employing formal logic in his proofs and demonstrations, and Ptolemy writing works on epistemology and even designing his *Optics* with the overriding purpose of resolving logical debates over the accessibility of empirical knowledge.[261]

The point of this digression is that when evaluating the different degrees of science education to which an ancient student of philosophy would be

260. For example, Cicero, *On the Boundaries of Good and Evil* 5.20.57; Maximus of Tyre, *Orations* 6, 10, 13, and 27; and it's a repeated theme in Seneca's *Natural Questions* and Galen's *On the Errors and Affections of the Soul.*

261. On this purpose behind Ptolemy's *Optics* see A. Smith 1999 and LeHoux 2012 (esp. 106–32); for his extant treatise on epistemology, see Huby & Neal 1989. Sextus the Pyrrhonist, author of *Against the Logicians* in two volumes (= *Against the Dogmatists* 1–2 = *Against the Professors* 7–8), was also a medical scientist of the Empiricist school: *OCD* 1358–59 (s.v. "Sextus Empiricus"). Hero, to give just one example, produces a formal proof of his theorem of least action to explain the laws of reflection in his *Katoptrics.* And Galen's *Institutio Logica* remains the only real textbook in formal logic to survive from the Roman period (on this and his other writings on logic see Morison 2008).

exposed, we must account for the effect of medieval Christian selectivity in preserving texts from antiquity, which has skewed our perception of what was normal. For example, the assumption is sometimes made that Roman Stoicism shifted its focus almost entirely to ethics and more or less frowned on logic and natural philosophy as pedantic distractions. But this is an illusion created by a later Christian preference for preserving the treatises of Stoics who took that position (a position which the Christians largely shared, as we'll see in chapter nine), when really those authors (and usually named here are Epictetus and Seneca) were freaks in comparison with their peers. In truth, even the great moralist Seneca was an advocate of science education and the advancement of scientific knowledge,[262] making Epictetus a freak *even next to him*. And yet Epictetus was not so wholly opposed to a useful education in science and logic, either.[263] And he was at the furthest extreme, and thus as unrepresentative as anyone could be. As Jonathan Barnes concludes:

> His own apparently consuming interest in moral precepts and ethical improvement was not characteristic of the Stoic philosophers of his age. On the contrary, in this respect at least Epictetus was unorthodox: his contemporaries—Stoic teachers and Stoic pupils—were obsessed not by ethics but by logic; they gave themselves to logical matters with a passion, a single-mindedness, and no doubt a pedantry which galled Epictetus—as it had galled Seneca, and as it has later galled so many earnest philosophers. Doubtless Epictetus exaggerates. Nonetheless, it seems to me beyond doubt that logic engrossed men during this period in the history of philosophy as it has rarely engrossed men in any other period.

And this was no basic logic they were involved in, but a logic "more subtle, more advanced, more technical" than almost anything that would be seen until modern times.[264] Science was thoroughly embroiled in all of this. For just as Barnes found for logic, a Stoic passion for physics is likewise documented,[265] which has also been obscured by a subsequent Christian disinterest in preserving very many Stoic treatises in that field. Hence across

262. Evident throughout Seneca, *Natural Questions*. See Lehoux 2012: 77–105.

263. See Epictetus, *Discourses* 1.7, with Crivelli 2007 and Long 2002: 149–52.

264. J. Barnes 1997: 126.

265. See Barker & Goldstein 1984.

the empire Stoic education did not avoid scientific knowledge but was immersed in it. The same could be said, even if in differing degrees, of all the major schools, from the Platonists and Aristotelians to the Epicureans and Skeptics. Exposure to scientific knowledge and methods in their lectures and debates would have been considerable for any layperson of the time, even when varying from teacher to teacher.

After philosophy school came specialized science education. Many scientists would move from the philosophy schools into the tutelage of a scientific expert, although some might have attempted to find such an arrangement without much or any prior training in philosophy—or in some cases even the *enkyklios*, though such students would have to tutor under what we would call a quack. Such students would not become 'ancient scientists' in our sense of the term, but pseudoscientists, teachers of nonsense who did not use any valid methods or even heed the actual scientific knowledge the ancients had then achieved. There were a great many such quacks in the ancient world, far more so than today, since there were no regulatory agencies and not much popular understanding of how to tell a quack from a genuine expert. Everything we have surveyed so far on the rarity of education, and the even greater rarity of exposure to significant science content, or even training in critical thought, ensured the success of quackery and pseudoscience in all fields, and Galen attests to the widespread existence of quack teachers in all scientific subjects.[266] But there *were* genuine experts who sought to compete against this sea of quacks and pseudoscientists through their mastery of the skills of rhetoric in addition to the knowledge of their own particular art, and therefore there were serious schools all over the Roman world for prospective science students who wanted real knowledge.

As just one example of how genuine knowledge was taken seriously, Strabo comments on the need for sound facts in writing on geography:

> The person who attempts to write an account of the countries of the earth must take many of the physical and mathematical principles as

---

266. Galen, *To Thrasybulus* 22 (= Kühn 5.843); *On the Affections and Errors of the Soul* 2.3, 2.5 (= Kühn 5.69–71, 5.91); *On My Own Books* Kühn 19.9, 19.52; *That the Best Doctor Is Also a Philosopher* 2 (= Kühn 1.57); *On the Sects for Beginners* 6.14–15 (see Walzer & Frede 1985: 10–11 and Hankinson 1994: 1781–82). On quack engineers: Vitruvius, *On Architecture* 6.pr.6–7.

> hypotheses and elaborate his whole treatise with reference to their intent and authority. For, as I have already said, no architect or engineer would be competent even to fix the site of a house or a city properly if he had no conception beforehand of climates and celestial phenomena, and of geometrical figures and magnitudes, and heat and cold, and other such things—much less a person who would fix positions for the whole of the inhabited world.[267]

And this only reiterated what Strabo had argued from the start. As Serafina Cuomo observes, "Strabo opens his book by saying that the study of his subject, geography, requires extensive knowledge: philosophy, natural history, and especially astronomy and mathematics."[268] Thus, by building these requirements for establishing someone's work as 'genuine', a difference could be drawn between real and fake expertise, for an architect or engineer as much as for a geographer. And of course similar ideas are found throughout the works of Galen in the matter of medicine, and Vitruvius in the matter of engineering.[269]

I will discuss Galen and Vitruvius more specifically in a moment. But Galen certainly agreed with another Roman-period treatise (pseudonymously attributed to Soranus of Ephesus) which insists that medical students acquire competence in both natural philosophy specifically and the full *enkyklios* generally, while Vitruvius outright says so of engineering. This reflects a professional expectation that a doctor or an engineer should be, as Kudlien says, "a cultivated man who follows the standards of the higher social classes."[270] Thus we see Galen's treatise *That the Best Doctor Is Also a Philosopher* arguing the "eclectic" position that all branches of all the major schools of philosophy must be studied by a doctor, doing which would also have required prior completion of the *enkyklios*, including mathematics and astronomy.[271] Likewise, a pseudonymous letter of unknown date (but attributed to Hippocrates) insists that doctors learn mathematics, and

---

267. Strabo, *Geography* 2.5.1–2 (cf. 1.1.13).

268. Cuomo 2001: 178–80.

269. Most directly and completely in Vitruvius, *On Architecture* 1.1 and 6.pr.; and Galen, *On Examinations by Which the Best Physicians Are Recognized.*

270. Kudlien 1970: 20.

271. Ibid. 3–4 (= Kühn 1.60–63); mathematics and astronomy: ibid. 1 (= Kühn 1.53–54).

actually describes the practical use of mathematical knowledge by a doctor: geometry was necessary for understanding the mechanics of joints and bone structure, especially as an aid to surgery, and arithmetic was necessary for carefully analyzing the courses of diseases and treatments (and we can be sure it was equally essential for the measurements and conversions required in a doctor's apothecary art).

Though a case for doctors to study astronomy was more difficult to make, it was still part of the *enkyklios*, so the best doctors would be expected to have a basic competence in it. And in fact, Galen argued that determining seasons and latitudes (which required astronomical knowledge), and being able to predict the course of the sun throughout the day (likewise), were essential to medicine, and therefore so was astronomy.[272] Comparably, Vitruvius argues throughout his treatise *On Architecture* that an engineer should study all branches of science and philosophy, and that this breadth of study in fact would distinguish a real engineer from a hack, setting up the same kinds of standards in his field as Galen and others advanced for their own.[273] As noted before, the same purpose with respect to the medical profession is obvious in Galen's treatise *On Examinations by Which the Best Physicians Are Recognized*. Even the best midwives were expected to be literate and ideally even more educated than that.[274] Even as far back as the fourth century B.C. astronomers and surveyors were expected to be educated.[275] The architect of the Mausoleum (commonly known as one of the "seven wonders" of the ancient world) wrote a book arguing this very point in the same century, which advocated a full liberal arts education for engineers.[276]

Likewise, several passages in the *Categories of Fields* by Hyginus Gromaticus (written in the second century A.D.) demonstrate that

---

272. Argued in Galen's *Commentary on Hippocrates' 'Airs, Waters and Places'*. Other uses for astronomy to ancient doctors are surveyed in Hulskamp 2012.

273. For example, Vitruvius, *On Architecture* 1.1.4 and 6.pr.5–7 (see also Rowland & Howe 1999: 13; Goguey 1978; and Galen, *On the Affections and Errors of the Soul* 2.3 = Kühn 5.68–69).

274. Soranus, *Gynecology* 1.3–4.

275. Xenophon, *Memoirs* 4.2.10.

276. For sources and discussion on the contents of this lost work see *OCD* 1247 (s.v. "Pythius") and *EANS* 712 (s.v. "Putheos of Priene").

surveyors were expected to know astronomy, and that Hyginus himself was well read in the sciences. Many other surveying treatises from the early empire show that surveyors and engineers were expected to be both educated and excellent mathematicians.[277] Writing in middle of the first century A.D., Columella says surveyors and mathematicians ("measurers and calculators") had their own teachers and educational traditions just as orators do, and says there were actual schools for them.[278] He goes on to argue that even agriculturalists should be educated in the subject of natural philosophy, especially astronomy and climatology, and he declares the requisite knowledge can only come from formal education.[279]

Though Columella is arguing *for* the formalization of agriculture as a science, and not describing what was actually the case, it is relevant that Columella understood that agriculture as an academic field could only gain social respect and make real progress if it were formalized with an underlying theory and its practitioners were fully cultured and educated in all the basic arts of the *enkyklios* as well as philosophy, just as Strabo had said of geographers, and Galen of doctors, and Vitruvius of engineers. The most respected professions were thus often held to certain standards that made it necessary for a serious and ambitious doctor, engineer, architect, or astronomer to complete a well-rounded education that touched upon the whole range of sciences. The best professionals had to be ready to demonstrate not just that their skills and methods worked, but that they were serious men of culture and that their art was based on a defensible and sophisticated theory.[280] The most successful way to do that was to study science and natural philosophy and establish one's art on the principles established therein. Consequently, science education for actual scientists remained strong, at least into the third century A.D.

So what exactly did that education consist of? Study under professors of the sciences emphasized oral and practical instruction over book learning.[281]

277. Cuomo 2001: 170–73.

278. Columella, *On Agricultural Matters* 1.pr.3, 1.pr.5.

279. Columella, *On Agricultural Matters* 1.pr.22–24 (and cf. 1.pr.32).

280. As argued, for example, in Barton 1994a, Pearcy 1993, and von Staden 1997.

281. Cribiore 2001: 145–46; on all fields in the Roman period see "Professional Education" in Clarke 1971: 109–18. For a school of 'Egyptian' medicine before the Roman period, which might have established a model for later scientific schools,

But this is not entirely different from today, where labs, practicals, lectures, and internships are viewed as essential, for example, to a doctor's education and even for many engineers and astronomers. Likewise, as today, textbooks and reading were still an essential part of ancient medical education, and even more so for the mathematical sciences, while lectures and demonstrations were just as effective (if not more so) in transmitting scientific knowledge. In fact, any given student would become familiar with several different kinds of written texts in their studies, including poems, lecture notes, treatises, collections, dictionaries and encyclopedias, "problems and questions" (similar in function to what we now call workbooks and FAQs), letters, introductions (what we now call textbooks), and commentaries.[282] It's unlikely a typical student would (or even could) buy all of these. They most likely used copies owned by their teacher or available at a local city library; and of course they would have shared each other's texts, as much as they owned any.

Although it was common in antiquity (though not always the case) for a scientist's sons or kinsmen to follow the same occupation—and certainly kinsmen had an advantage over other students, as their education would be considerably cheaper overall and easier to secure—the evidence is clear that most students of the sciences came from outside a scientist's family.[283] They would attend a scientist's lectures, read books on the subject, and often observe or work for their professor in the role of an apprentice. As I've mentioned before, there were really only two scientific 'careers' that one could make a living from: medicine and engineering. These broad categories encompassed numerous specializations and separate interests for any given individual to master or pursue (like pharmacology or surveying), but generally all life sciences were undertaken by doctors and all physical sciences by engineers. The only thing close to being an exception is astronomy, which besides being an occupation of engineers, was also seriously pursued

---

see Cribiore 2001: 25.

282. On the different types of texts used in ancient science education and their relation to oral instruction see Taub 2008: 13–29, with Doody et al. 2012, Netz 2011, Nicholls 2010, Doody 2009, Asper 2007, Horsfall 1979: 81–82, and Witty 1974.

283. Galen, *On Conducting Anatomical Investigations* 2.1 (= Kühn 2.280–83); Vitruvius, *On Architecture* 6.pr.4–7.

and advanced by many astrologers.[284] Although even these occupations could overlap (the Roman astronomer Ptolemy, of the mid-second century A.D., clearly mastered both astrology and engineering), it was at least possible to make a good living as only an astrologer, and so the existence of a demand and social position for astrologers will have generated some scientific astronomers (as well as a lot of quacks).[285] But in most cases, it was the career of doctor or engineer that produced the accomplished scientists of the Roman era.

We know quite a lot about the education of doctors, but very little about the education of engineers.[286] Like doctors, aspiring engineers will have relied more on hands-on work as an apprentice than on books and lectures, but their education must still have included a significant amount of the

---

284. On the nature and content of astronomy education in antiquity see Evans & Berggren 2006. On mathematics education: Cuomo 2001 (with Cuomo 2000, Evans & Berggren 2006: 43–48, 243–49, and other references I cited on mathematics education in chapter five).

285. Astrology was an art taken quite seriously at the time, although not by everyone: see Barton 1994b (and 1994a) and *OCD* 187–88 (s.v. "astrology"); for ancient arguments pro and con see Long 1982 and Sextus Empiricus, *Against the Professors* 5. All Epicureans and Skeptics rejected astrology, but so did others; even the occasionally gullible Pliny the Elder: cf. *Natural History* 2.6.28–29. Nevertheless, astrology was not only lucrative and popular, it also typically demanded real scientific expertise in astronomy (on which besides Barton, see A. Jones 1994). On distinctions made between astronomy and astrology: Losev 2012.

286. On medical education in antiquity (to which I shall soon turn) see: Kudlien 1970; Clarke 1971: 109–12; Nutton 1975 and 1995 (with 1993: 11–15); Kollesch 1979; Duffy 1984; Todd 1984; Iskandar 1976 and 1988; Jackson 1988: 58–64, 129–30; and Kleijwegt 1991: 135–63. Together, these authors (and the scholarship they cite) also demonstrate (among other things) the existence of formal medical associations in dozens of cities throughout the Roman empire (on which also Korpela 1987: 102–06), as well as several legionary hospitals (*valetudinaria*)—and wherever doctors associated and worked in significant numbers, there would have been students. For more on Roman military hospitals, some of which were the most advanced medical facilities in the world until early modern times, see: Scarborough 1968; Davies 1970; Harig 1971; Pitts & St. Joseph 1985: 91–103; Korpela 1987: 106–10; Press 1988; Jackson 1988: 65, 113, 133–37 and 1993: 88–89; Wilmanns 1995; James & Thorpe 1994: 6; Nutton 2004: 178–82; Peters 2010. On the (perhaps less) scientific use of Asclepian temples as early civilian hospices, see discussion and sources in P. Green 1990: 487–89 and Nutton 2004: 103–10.

latter. In fact, much of what one might say about medical education probably applied to an engineering education as well, even if there were differences in degree. Marrou's claim that only medicine managed to develop its own particular type of training is not believable, especially given the evidence in Vitruvius and the treatises of other Roman surveyors (the *agrimensores*) and engineers.[287] In fact Peter Rosumek found so much consistency in Roman mining operations across the empire he concludes the Romans must have had engineering schools with an established body of texts or instruction.[288] There is evidence supporting the same conclusion from other engineering endeavors, such as hydrological engineering (in not only but especially aqueducts) and pump design and manufacturing.[289] And just as with doctors, the legions employed so many engineers there must certainly have been students following on, acquiring and then disseminating standard practices and principles.[290] In fact, if anything, medical schools must have been even *more* varied and diverse in their procedures, emphases, and curricula than engineering schools would have been—for unlike medicine, there were no sectarian divisions among engineers, as there was very little room for disagreement about what they had to know or be able to do, or even how to do it. Because the placebo effect will not hurl bullets or hold up a roof.

The fact that we know a lot more about medical education in antiquity than for any other science should not be taken to imply a difference in social value, but rather a difference in practical demand. More books by or about doctors have survived than books by or about engineers or astronomers—and in fact, a vast deal of what we know about medical education comes from the survival of dozens of books from a single and uncommonly chatty author: Galen. So the literary record has also been skewed in favor of medicine. More importantly, however, and perhaps part cause of the disparity in the literary record, there would have been far more doctors than scientists of any other

---

287. Marrou 1964: 287–91 (= Marrou 1956: 191–94). We actually know more about engineering education than Marrou is aware: see Donderer 1996: 57–62 (and 70), Goguey 1978, Dilke 1971: 47–65, and (in general) the discussion of surveyors and engineers throughout Cuomo 2000 and 2001, and the introductions to Rowland & Howe 1999 and DeVoto 1996.

288. Rosumek 1982, cf. p. 165.

289. See Oleson 2004 and Greene 1992

290. Cf. e.g. E. Evans 1994.

kind. As a matter of practical economics, doctors must always have been in far greater demand, and therefore schools, associations, and professors of medicine would have been more numerous (and the physical evidence bears this out). For example, there must always have been hundreds of times more people, even among the elite, suffering from or worrying about illness or injury (whether real or imagined) than undertaking major building projects, while even in the army, more doctors would be needed to tend to the troops than engineers would be needed to tend to the artillery and military construction projects. Scientific astronomers would be even rarer than engineers—unless the demand for sophisticated 'scientific' astrologers was enormous, which is unlikely. This is not to say any of these professionals would have been hard to find, at least in urban areas, but only that scientific doctors would have greatly outnumbered experts in other fields.

To close my discussion of science in higher education I shall provide two examples, one in each major scientific field, Galen for medicine and Vitruvius for engineering, the only scientists whose education we know much about. Though they are in some respects exceptional, their stories probably reflect the experience of the best scientists of antiquity, while the education of lesser lights would still have followed similar patterns even if to a lesser degree.

Galen's father, an accomplished engineer, taught his son arithmetic, geometry and even more advanced mathematics (such as rudiments of trigonometry and mechanics), as well as grammar and "other subjects," instilling in Galen a lasting respect for mathematics and mathematical sciences, and their methods.[291] Thanks to his father, Galen had completed the

291. Galen, *On the Affections and Errors of the Soul* 1.8 (= Kühn 5.41–42); *On My Own Books* Kühn 19.39–43 (where Galen also says his father learned mathematical subjects from his grandfather and great-grandfather, suggesting a family tradition in the engineering profession). Supporting the inclusion in Galen's education of rudiments of trigonometry (spherics and conics, including some knowledge of the production of conical sundials), see Galen, *On the Affections and Errors of the Soul* 2.1 (= Kühn 5.59–60). On Galen's use and knowledge of mathematics and mathematical sciences in his works and methodologies in general see Lloyd 2005. For important examples see: Galen, *On My Own Books* 11 (= Kühn 19.40), *On the Affections and Errors of the Soul* 2.3–7 (= Kühn 5.66–103), *On Treatment by Venesection* 3 (= Kühn 11.255–56), *On the Doctrines of Hippocrates and Plato* 8.1.19–21 (with 9.4.30–31), *On the Therapeutic Method* 1.4.4–6 (with 1.4.12 and 1.5.1, and related notes in Iskandar 1988: 158 (§P.68,14–15)). For examples of Galen's astronomical

entire *enkyklios* (except rhetoric and dialectic) by the age of 15. At that point his father sent him to study dialectic and then philosophy under various professors, first under a Stoic and a Platonist, then an Aristotelian, then an Epicurean—all of whom were available in his home city of Pergamum. His father probably intended him to become an engineer (it seems to have been a family trade going back many generations), but he had a dream in which a god urged him to educate his son as a doctor. So at age 17 Galen began studying medicine alongside philosophy. After his philosophical and basic medical education were reasonably complete, Galen pursued more advanced medical studies under numerous teachers in several cities over many years, wherever he could learn something new.[292] Though he never mentions having attended any school of rhetoric, he must have mastered the fundamental elements of rhetoric under his philosophy professors, for his abilities as a writer and speaker are evident throughout his works and in many of his autobiographical digressions. The Stoics and Aristotelians in particular valued rhetoric as part of their philosophical curriculum, and Galen studied under both. Galen was always extremely grateful for his broad mathematical and scientific education and often praised it as the ideal for others to follow, although he knew only the fortunate would be likely to so fully as he. Galen also came from modest wealth—although originating from the professional working class, and not from "old money" aristocracy. His father was a retired engineer, who acquired a few small estates, and eventually entered local politics, and had done well enough in life to leave a relatively small inheritance that Galen found enough to live on.[293]

Near the end of his life Galen wrote a survey of his extensive writings for the purpose of articulating his own ideal curriculum for medical school,

---

knowledge and interests see Strohmaier 1993 and especially Galen's *Commentary on Hippocrates' 'Airs, Waters and Places'*. For his knowledge and interest in these, and other sciences as well, see Nutton 1999: 169–70 (§P.82,19).

292. For more detail on Galen's education: *DSB* 5.227–29 (in s.v. "Galen"), *NDSB* 3.91–93 (in s.v. "Galen"), Hankinson 1991: xix-xxii and Nutton 2004: 216–19 (and Nutton 1973 establishes the chronology of Galen's education and early career). See also *EANS* 335–39 (s.v. "Galen of Pergamon"), Hankinson 2008, and Mattern 2013: 36–80.

293. Galen, *On the Affections and Errors of the Soul* 1.9 (= Kühn 5.48). For evidence regarding Galen's inheritance see Iskandar 1988: 145 (§P.42,12) and Nutton 2004: 389 (notes 4 and 11) with Hankinson 2008: 355–90.

which started with logic and principles of method, then placed great emphasis on autopsy, vivisection, and other forms of hands-on experience.[294] In terms of order, emphasis, and subject material, Galen's curriculum was a selection of the best from the standards employed by many teachers of his day and before, though presented in his own words. Marrou's claim that medical education had ceased to employ much dissection in the Roman period until Galen 'revived' the practice, for example, is indisputably false.[295] Galen's many treatises on anatomy leave no doubt that dissections of animals as proxies for humans were widely employed in medical schools throughout the empire as aids to teaching, educational observation of human surgery and public anatomical demonstrations were common, and anatomical research already thrived in the Roman period before Galen.[296] As much as a century earlier Plutarch had already mentioned public 'surgical' demonstrations by doctors as a common method of attracting patients.[297] Around the same time Dio Chrysostom described public reactions to anatomical demonstrations, which were clearly already well in vogue.[298] And we know public medical 'contests' were held in Ephesus and possibly Smyrna, and likely elsewhere.[299]

Galen lived and wrote in the late second and early third century A.D. The engineer Vitruvius lived and wrote two hundred years before him.[300] But the

---

294. Galen, *On My Own Books* (especially = Kühn 19.52–61). See also Hankinson 1994: 1782–84.

295. Marrou 1964: 290 (= Marrou 1956: 193).

296. Galen's *On My Own Books* is full of references to public anatomical and surgical demonstrations. See also Galen, *On the Uses of the Parts* 15.1 (= May 1968: 658) and *On Examinations by Which the Best Physicians Are Recognized* 9.6.

297. Plutarch, *How to Tell a Flatterer from a Friend* 32 (= *Moralia* 71a).

298. Dio Chrysostom, *Discourses* 33.6.

299. See Nutton 2004: 250. For a full survey of the evidence for scientific dissection and public medical, anatomical and surgical demonstrations, lectures, and contests in the early Roman empire see: Kudlien 1970: 20–21; Ferngren 1982: 278–79; Nutton 1995 (with 1985: 27); Debru 1995; von Staden 1995 and 1997; Byl 1997; Selinger 1999; Rocca 2003: 1–14; and Mattern 2008: 69–79. Relevant material can also be found in: Singer 1956; Duckworth, Lyons and Towers 1962; May 1968; Nutton 1971a; von Staden 1975; Lloyd 1983; Furley & Wilkie 1984; etc. I will discuss this 'revival' of dissection under the Romans before Galen (and the occasional practice of human dissection) in *The Scientist in the Early Roman Empire*.

300. For more on Vitruvius see *DSB* 15.514–21 (s.v. "Vitruvius Pollio"), *OCD*

two are connected through Galen's father and the education in engineering he gave him—in fact, it's possible Galen had read and admired Vitruvius, and he certainly shared many of the same values.[301] Vitruvius describes his own education in his preface to the sixth book of *On Architecture*:

> I thank my parents immeasurably and bear them great and infinite gratitude because in accordance with the spirit of the Athenian law they had me trained in an art, an art, moreover, that cannot be mastered without education in letters and comprehensive learning in every field. When, therefore, I had a stock of knowledge increased both by the solicitude of my parents and the erudition of my teachers, and enjoyed myself by reading both literary and technical writing, I stored up all these assets in my mind.[302]

Hence Vitruvius refers to having had several teachers and to reading both literary and technical works.[303] The "Athenian law" he refers to held that parents have the right to compel their children to care for them in old age *unless* they failed to educate their children in some art.[304] For Vitruvius this showed a correct concern for the great value of learning a professional skill.

Later Vitruvius says it used to be the case that architects "never trained

---

1561–62 (s.v. "Vitruvius (Pol(l)io)"), and *EANS* 830–32 (s.v. "M. Vitruuius Pollio").

301. Galen's *Exhortation to Study the Arts* contains enough uncanny coincidences with remarks in Vitruvius' *On Architecture* that Galen must have read and liked it (or else some Greek work Vitruvius followed quite faithfully, as some scholars suggest he did). Compare, for example, *Exhortation* 5 and 8–9 (= Kühn 1.15, 1.20) with *On Architecture* 6.pr.1, 6.pr.4., and 9.pr.1–2.

302. Vitruvius, *On Architecture* 6.pr.4.

303. Vitruvius, *On Architecture* 6.pr.5 mentions again his having several teachers, who taught him professional ethics as well as the skills of his field, and 9.1.16 mentions his learning astronomy from several teachers.

304. Vitruvius, *On Architecture* 6.pr.3 (repeated in Galen, *Exhortation to Study the Arts* 8, = Kühn 1.15). For something of the underlying sentiment see Xenophon, *Economics* 20.15. Note that Vitruvius is not saying he was raised in Athens or even that this Athenian law was still in force; he is merely using this as a popular example of a social ideal that his parents fulfilled (an example of a literary detail that all his readers would know from having studied the same classical works in school as he did).

anyone but their own children and relatives," which entails that in his own time it was not unusual for engineers to take students from outside their family.[305] Galen says the same of the medical profession, and was himself an example of the fact.[306] Vitruvius also mentions that engineering was so profitable that many amateurs and hacks attempted to pursue jobs in the field, driving many employers to simply manage their own building projects rather than trust the work to unreliable and ignorant men.[307] Hence, following the general trend I've noted already, Vitruvius declares that one of his purposes in writing is to explain what a true (and hence reliable) architect should know, advancing a set of professional standards by which an employer can assess the quality of an engineer.[308] The education Vitruvius insists upon (and that clearly he had received himself) consisted of grammar, drawing, arithmetic, geometry, optics, history, philosophy (with an emphasis on *physiologia*, or natural philosophy), music (especially the mathematical principles of harmonics and acoustics), medicine, law, and astronomy, as well as of course all the principles of mechanics and building.[309]

Vitruvius explains that natural philosophy (which meant a broad knowledge of the ancient sciences and the philosophy thereof) was essential for understanding other writers on engineering and for understanding the physics and behavior of things like gravity and water.[310] The sciences associated with music were essential for designing the acoustics of theaters, and useful in calibrating catapults and crossbows, and in constructing water organs and "other hydraulic machines."[311] Astronomy was essential for constructing accurate sundials and for improving elements of building and city layout, while medicine was essential for adapting building and city layouts to climate and matters of hygiene, such as knowing that lead pipes

---

305. Vitruvius, *On Architecture* 6.pr.6.

306. Galen, *On Conducting Anatomical Investigations* 2.1 (= Kühn 2.280–83).

307. Vitruvius, *On Architecture* 6.pr.6–7.

308. Vitruvius, *On Architecture* 1.1. See Iskandar 1988.

309. Vitruvius, *On Architecture* 1.1.4–10 (mechanics is added in 10.pr.3).

310. Vitruvius, *On Architecture* 1.1.7.

311. Vitruvius, *On Architecture* 1.1.8–9 and 5.4–5.

are poisonous.[312] Arithmetic and geometry, of course, were needed for many reasons, including making accurate measurements and calculating costs.[313] Optics was also important, especially for visual design.[314]

Though Vitruvius does not explicitly mention rhetoric or dialectic, it is clear he studied these subjects and imagined other engineers would, too.[315] In short:

> To be educated, the engineer must be an experienced draftsman, well versed in geometry, familiar with history, a diligent student of philosophy, know music, have some acquaintance with medicine, understand the rulings of legal experts, and have a clear grasp of astronomy and the ways of heaven.[316]

His requirement of being facile with history and law would entail attending a school of rhetoric, as that was the only place such subjects would be taught. Thus we can be sure he completed a full course of higher education, rhetoric and philosophy, as well as the *enkyklios*, for we see the full quadrivium in his training requirements. Thus he imagined a very broad education, covering the whole *enkyklios* and beyond. He argues that all these branches of knowledge are interrelated, so learning them is not excessively difficult, especially since the engineer does not need complete mastery, only a reasonable competency in each.[317] Nevertheless, he was aware of at least a very few people who had mastered many arts in considerable depth, and he holds them in high esteem.[318]

Like Galen, Vitruvius emphasized the importance of combining

---

312. Vitruvius, *On Architecture* 1.1.10; on lead pipes: 8.6.10–11.

313. Vitruvius, *On Architecture* 1.1.4.

314. Vitruvius, *On Architecture* 6.2 and bks. 3, 4, and 6; on this point see also Athenaeus the Mechanic, *On War Machines* 28.5–12, along with Whitehead & Blyth 2004: 139–40.

315. Vitruvius, *On Architecture* 9.pr.17–18 (see Rowland & Howe 1999: 8).

316. Vitruvius, *On Architecture* 1.1.3 (cf. 1.1.13).

317. Vitruvius, *On Architecture* 1.1.11–18.

318. Vitruvius, *On Architecture* 1.1.17, where he says masters of all the mathematical (always physical) sciences were called "mathematicians" in honor of their intellectual accomplishment.

apprenticeship and practical hands-on experience with a formal education.[319] We have a confirmation of this a century later by the Roman engineer Hero, as paraphrased by Pappus in the mid-4th century:

> The engineers who follow Hero say mechanics has two parts, one theoretical and the other practical, and that the theoretical part is assembled from geometry, arithmetic, astronomy, and natural theory, and the practical part from metalworking, building, carpentry, painting, and training in these subjects by hand. And so, they say, one who develops in these skills and sciences from childhood, and has a natural talent for them, will be the greatest engineer and inventor of mechanical devices.[320]

The word used here for the "practical" part of mechanics is *cheirourgikon*, literally "hands-on work" or "work done by hand." The word used for the "theoretical" part is *logikon*, which is said to consist not only of physical theories (*tôn phusikôn logôn*) but also the mathematical study of geometry (*geômetria*), arithmetic (*arithmêtikê*), and astronomy (*astronomia*), and we can assume Hero would also have included harmonics, as Vitruvius says that was commonly expected of any engineer, and surely no one could imagine himself "the greatest engineer and inventor of mechanical devices" without a command of so popular and fundamental a field. Since that completes the quadrivium, and since the trivium (grammar, rhetoric, and logic) was a fundamental preparation for the study of all advanced mathematics and physical theory, Hero's school was essentially requiring a complete formal education, just like Vitruvius.

Hero's school was based at Alexandria (hence "Hero of Alexandria"), but just as an advanced medical education could be received in many other major cities in the Empire (as Galen attests), an engineering education surely would have as well—professional engineering being of even greater importance to national security and imperial urban economies (and not only in the construction of aqueducts). We have hints confirming

319. Vitruvius, *On Architecture* 1.1.1–2, 1.1.11.

320. Pappus, *Mathematical Collection* 8.1.(1024). This probably paraphrases a lost work by Hero; and in accord with the principles expounded here, all of Hero's extant treatises are explicitly written as instructional textbooks, and typically include a survey of the underlying science as well as practical advice on building machines.

this. Strabo remarks, for example, that near the turn of the common era Marseilles was still famous for the brilliance and industry of its engineers, as "traces of their ancient zeal are still left among the people there, especially in regard to the making of instruments and to the equipment of ships," and though Strabo complains that they have taken more now to philosophy and oratory, that remark actually makes the existence of formal schools attended by engineers in Roman Marseilles all but certain.[321] There is no way it could maintain a peculiar reputation for its engineers otherwise. That other centers of learning existed for engineers is strongly suggested by the evidence (noted earlier) of standard schooling in mining operations and other routine endeavors empire wide (like urban building and pump and aqueduct design).

An essay by Edgar Zilsel published at the end of WWII is still often cited as declaring there was a "wall which since antiquity had separated the 'liberal' from the 'mechanical' arts," but Hero and Vitruvius alone refute this.[322] That wall was breached at precisely their level: the well-to-do middle class, or "professional class" as it were, which stood well above the shopkeepers and craftworkers in social station, though still below the landed aristocracy—those lifelong "men of leisure" who, unlike Vitruvius, never had to work for a living (on my categorization of ancient social classes, see note in chapter two). Everything we have surveyed so far about ancient education establishes beyond doubt that it was impossible to be as literate, eloquent, and well-read as Vitruvius or Ptolemy or Hero without a substantial education in the liberal arts. As Vitruvius explains, and in his own writing reveals, quality engineers like him were expected to be literate and well-educated. Yet there were no special grammar schools for engineers. To become an engineer you had to complete a standard primary and secondary literary education, and to be very successful, you also had to complete the same *enkyklios* as anyone else might, and study under the same philosophy professors, and master the same principles of rhetoric and dialectic. Every other Greek and Roman writer on engineering or physical science (likewise medical or any

---

321. Strabo, *Geography* 4.1.5.

322. Zilsel 1945: 342. Ironically Zilsel was probably reacting to the *modern* British educational system, and projecting its impractical classics-oriented studies onto antiquity (see quotation of McGrayne 2011: 63, and my related discussion and note in chapter one).

other science) confirms the same conclusion: all must have completed some higher education in rhetoric, since it was only there that any real skill and practice with prose composition began. Though medicine and engineering were specialized fields, pursued by a very select few, there is no indication that those who pursued them were separated by any 'wall' from others among the educated elite in literary or philosophical background. To the contrary, one of the very reasons that leading professionals insisted on a broad, well-rounded education for members of their craft is that it pulled them into communion with the social elite, ensuring that a professional's education substantially overlapped with that of the wealthy and powerful, the very people those professionals would constantly have to court and impress.

It seems unlikely that these standards for the education of philosophers and scientists continued in the Middle Ages, which oversaw a broad decline of scientific knowledge, and the gradual elimination of even the idea of a philosophy school. Cathedral schools and private teaching went on, but whether they could claim comparable science content is doubtful. Even with the rise of the universities, it is hard to see even their science curriculum as comparable, much less surpassing, what ancient Roman philosophers and scientists had access to. But whether that's the case I must again leave for others to explore. Although I shall demonstrate some of these points by comparison myself in *The Scientist in the Early Roman Empire*.

8. State & Public Support for Education

Aristotle insisted "it is clear there should be legislation about education and it should be conducted on a public system" that was the same for all citizens, because (he argued) neglecting this leads to social and political decline.[323] But Aristotle's dream was never realized. The most ever done was to create subsidies or salaried positions for teachers, but never to explicitly control or dictate, much less normalize or improve the curriculum. As Morgan says, "notably absent is any official interest in what exactly teachers taught, or how, or to whom, or when, where, or to what degree, let alone anything so deterministic as their qualifications for teaching."[324] That is an exaggeration, since there were unwritten standards when selecting whom to appoint to a public teaching post, or when parents and students chose which teachers to hire or study under, and in practice, tradition and cultural expectation had a strong standardizing effect on school curricula. But the direction this took was never subject to political control.

Aristotle's notion of all free boys receiving the same education was a Greek democratic ideal partly fulfilled by private or municipal philanthropic foundations in various Greek cities over the centuries, and later in many Roman cities, but it was never universally realized.[325] The most common

323. Aristotle, *Politics* 8.1.1337a. Plato's *Republic* and *Laws* articulate similar claims.

324. Morgan 1998: 27.

325. For some social and cultural analysis of this trend, drawing on inscriptions and literary sources, see Nilsson 1955. As just a few examples, inscriptions attest

foundations only educated a limited segment of the free male population anyway (and sometimes girls), and did not always offer the best education. Marcus Aurelius was grateful to have learned "not to have attended the state schools and to have been provided instead with good teachers at home, and to know that one must spend lavishly on such things," and Pliny the Younger was also aware of a troubling quality problem in state schools.[326] But then, the same complaints are still heard about public schools today.

Though many such foundations might not have survived the looting or devastation wrought by Roman conquerors in the second and first centuries B.C., or the social and economic devastation of the Roman empire in the crisis of the third century A.D., many survived or were also established (or renewed) during the very centuries in between that marked the Pax Romana. As a remarkable example, Plutarch attests to a continued tradition of a full course of liberal arts (meaning the *enkyklios*) still being subsidized for citizen boys at Athens in the late first century A.D., which practice we know had continued for centuries.[327] Though this may have been exclusive and somewhat superficial, it definitely included studies in mathematics, grammar, rhetoric, and philosophy, and possibly other subjects as well, while its students copied out new books for the school library every year.[328] This was probably not unique. Similar ephebic colleges appear to have survived in many other Greek cities of the Roman era.[329] And besides these, "public" grammar schools were even more common.

Notably, all these municipal schools were local benefactions, either the

---

educational foundations for the citizens of Xanthus (*SEG* 30 [1980] no. 1535.24–28) and Teos and Miletus (*SIG* 2.577–78).

326. Marcus Aurelius, *Meditations* 1.4: i.e. *dêmosias diatribas* would normally mean "state-managed schools" rather than merely "common schools" (cf. *LSG* 387, s.v. "*dêmosios*" with ibid., suppl. p. 86); Pliny the Younger, *Letters* 4.13.6–8.

327. See Marrou 1964: 265 (= 1956: 176), with supporting inscriptions extending well into the Roman era: 1964: 280–84 (= 1956: 186–87), and commentary: 1964: 567–68 notes 1–2 (= 1956: 406); for more extensive discussion see Nilsson 1955: 21–29, Chankowski 2004, and Watts 2006: 24–47, and summary and bibliography in König 2009: 395. See also *OCD* 508, s.v. "*ephêboi*."

328. Tod 1957: 137, 139.

329. See Hin 2007; König 2005: 47–63; Kah & Scholz 2004: 104–24, 193–210; and Kleijwegt 1991: 91–101, 155. On the pre-Roman history of the function and state sponsorship of the *ephebeia*: Kozak 2013 and Casey 2014.

bequest of a wealthy philanthropist to a favored city, or the exceptional act of a city council, and never enacted on a wider scale. Even emperors only went so far as to endow salaried posts for professors of higher education in a few privileged cities. Thus many, perhaps most cities did not have such endowments, while people living outside major cities would almost always be left out. Moreover, most of these foundations, and possibly all of them in the Roman period, only subsidized secondary or higher education, leaving parents to provide or fund elementary education entirely on their own.[330] Even peripheral state support, such as exempting all teachers from taxes and levies and other civic burdens, which became standard throughout the early Roman empire, again only applied to teachers at the secondary and higher levels.[331] This would have had the effect of all but shutting out the children of the illiterate poor, in effect preventing significant expansion of literacy except for those who acquired enough financial success to 'move up in the world' and thus afford elementary school for their kids or buy a literate slave who could provide it—although outright *buying* an elementary teacher was certainly far more expensive than renting one, so that option was generally only available to the rich.[332] The *literate* poor, of course, could often provide their own elementary education to their kids, but this only perpetuated literacy within literate families, it did not expand their number (unless by operation of natural selection, if literate children had a higher differential reproductive success than non-literate—but that, even if it happened, would be extremely slow in disseminating literacy). In any case, by subsidizing only secondary and higher education, even the most fortunate communities were only serving an already privileged minority of their citizens. However, that would still mean a much larger proportion of the population would be educated than in societies without such amenities, so this phenomenon must be made an important element of comparison with the Middle Ages.

330. On all these facts: Harris 1989: 130–33, 141–44, 283, 307; Cribiore 2001: 63–64; Cuomo 2001: 30–32, 34–37, 39–40, 43–44; Clarke 1971: 8; Marrou 1964: 431–39 (= Marrou 1956: 301–08); P.J. Parsons 1976 (esp. pp. 410–14 and Appendix II: 441–46). On possible motives for the disparity in support between secondary and elementary education, see Christes 1988. One indirect exception may be certain charities that subsidized living expenses (discussed in chapter two), which could have made primary education more affordable to thousands.

331. Harris 1989: 235–36, 241–47; Clarke 1971: 8–9.

332. See Harris 1989: 258–59.

Focusing on what benefitted science education specifically, the first boon came when Julius Caesar granted citizenship to all doctors and professors of the liberal arts residing in Rome,[333] and as we've seen, the 'liberal arts' included the sciences of harmonics and astronomy, and geometry, which included mechanics and optics. The Roman citizenship was probably often easier to obtain for professors of the arts then and thereafter, and that conferred numerous advantages, not solely in regard to status. Scientists thus gained an important leg up at this transition point in history. The public also indirectly aided science education in another way, by the building and maintaining of libraries, lecture halls, and faculty dining clubs—whether by the state or private philanthropists, either way a common sight in cities across the empire (as we'll soon see). Another significant boon came in the form of public salaries, and certain tax and service exemptions granted to professors.[334] These did not single out scientists or natural philosophers, of course, but were focused more on physicians and professors of rhetoric. Although certain other teachers and professors also benefitted, the emphasis we have already seen on rhetoric and the trivium is further reflected in the focus of these state benefactions and exemptions. Only doctors received the same or greater attention, but although exempted and salaried doctors would likely have taken students as well, they were usually selected for these privileges to provide medical care, not medical education or research, and so the social values represented by these benefactions were not *aimed* at scientific education or research, although they would have benefitted both. Public salaries for doctors (among other benefits) had begun in Hellenistic cities even before Roman times.[335] Many engineers were also directly

---

333. Suetonius, *The Divine Julius* 42.

334. See for example Lewis & Reinhold 1990: 2.206–08 (§56). Nutton argues the tax and other exemptions for doctors can be dated as far back as Julius Caesar (cf. Nutton 1985: 29, 2000a: 964 n. 63, and 2004: 249–50), although Cassius Dio, *Roman History* 53.30.3 places their origin under Augustus. Imperial privileges awarded to professors are discussed in Marrou 1964: 440–43, Bowersock 1969: 30–42, Nutton 1971b, Cuomo 2000: 31–37, and Perrin-Saminadayar 2004. The whole of *Digest of Justinian* 27.1.6.1–12 documents that doctors, rhetors, philosophers, grammarians, and law professors were all granted exemptions at least as early as the mid-second century A.D., while *Digest of Justinian* 50.4.18.30 suggests they existed as early as Vespasian (in the 70s A.D.).

335. Cohn-Haft 1956; Meunier 1997; and Nutton 1977, 1981, 1985: 34, 2004:

employed by the state, and though there is no evidence of their receiving privileges, state engineers, like state doctors, would likely have taken apprentices just as private engineers did, thus by this avenue as well the state was indirectly subsidizing science education.[336] In much the same way, the formation of legionary hospitals under the early empire (see note in chapter seven) also added another prop to medical education by the widespread employment of military doctors and the providing of facilities where the need and opportunities for students would be considerable. And on top of all this, some public doctors *were* paid to teach.[337]

Vespasian also gave a major prop to medical education by granting doctors the right to form medical associations later in the 1st century A.D. This law also included professors of other subjects, granting them freedom of assembly, exemption from taxes, and special protections against local authorities, even ordering crimes against them to be more severely punished.[338] Doctors and teachers of the liberal arts (including rhetors, grammarians, and geometers) also received special attention from the government to ensure they were not cheated out of their fees.[339] Beyond that, however, the few benefits directed at philosophy professors, who were less common than doctors or professors of rhetoric, were not bestowed for their role or work as natural philosophers, but for the entire program of their

153–55.

336. See Vitruvius, *On Architecture* 10.16.3, who describes the Hellenistic office of state engineer at Rhodes. It's certainly well known that the Roman imperial government also maintained the regular employment of engineers, both military *and* civilian (e.g. Cuomo 2001: 158–59, 176).

337. See *Digest of Justinian* 27.1.6.9.

338. Korpela 1987: 102–06; Kollesch 1979: 512–13 (see also Woodside 1942: 128), with primary sources in *Fontes Iuris Romani Antejustiniani* 1.1 (1941): 230 (§77). However, I do not believe the infamous *Augustan History* when it says Alexander Severus "paid regular salaries and provided lecture halls to rhetors, grammarians, doctors, soothsayers, mathematicians, mechanics, and architects, and ordered subsidies for their poor and freeborn students" ('Aelius Lampridius', *Life of Severus Alexander* 44.4) since, besides this statement being uncorroborated and *prima facie* absurd, modern scholars generally regard both the author and the content of this biography to be fictional (although, as we'll see, Marcus Aurelius might have done something close to this).

339. *Digest of Justinian* 50.13.1.1–11.

teachings, which was always primarily ethical and logical. And when they did receive a state salary, they were not being paid to conduct research, but to teach. Thus, these benefactions did benefit science education somewhat, but often indirectly, and of course only at the level of higher education, and only in major cities, and still only for a privileged few.

In fact, in the mid-second century A.D., emperor Pius put a cap on the number of doctors, grammarians, and rhetors who could claim state exemptions, and the limit he set was ten doctors (plus five rhetors and five grammarians) in provincial capitals, seven doctors (plus four rhetors and four grammarians) in other cities with Roman law courts, and five doctors (plus three rhetors and three grammarians) in all other cities.[340] This did not mean cities always filled their quotas, or that cities could not have more doctors, rhetors, and grammarians. It only limited the number who could claim imperial privileges. This might have had the effect of motivating many doctors, rhetors, and grammarians to spread out into cities where slots for privileges were still available, thus making their services more geographically accessible. And since city councils had to decide whom to award their limited slots to (and the law specifically *required* those selected to be diligent in their work) this should also have had the effect of elevating the quality of the privileged practitioners.[341] But either way, the number of these slots was relatively small, and thus the benefit of these exemptions to the public was limited. Nevertheless, Galen lists the exemptions as one of the reasons people pursue a medical career.[342] And the limits set by Pius only pertained to imperial privileges—it did not hinder individual cities from establishing their own salaried positions for public doctors or teachers, and several did. As for philosophers, Pius did not fix a limit on the number of professors who could claim privileges because (or so he suggests) real philosophers would not ask for them, and philosophy professors were comparatively rare.[343]

340. Marrou 1964: 434–36 (= Marrou 1956: 301–03) and *Digest of Justinian* 27.1.6.2, reporting an interpretation of the third century Roman jurist Herennius Modestinus of the second century decision of emperor Antoninus Pius.

341. *Digest of Justinian* 27.1.6.4. On all the above aspects of Roman imperial support for doctors see Jackson 1993: 80–84 and Scarborough 1970: 297.

342. Galen, *On the Doctrines of Hippocrates and Plato* 9.4.3–6.

343. *Digest of Justinian* 27.1.6.7 (in the context of 27.1.6.5–9), which also suggests philosophers were expected to teach for free, or at least not to complain if their

Philosophers might only have been exempted from duties and burdens *other than* mere expenditures of money, since it was assumed that no philosopher could value money, and poverty was already grounds for exemption.[344] But we know philosophers did receive privileges and exemptions even from monetary obligations, so the backhanded remarks of Pius seem not to have affected their legal standing to claim them.[345]

Vespasian appears to be the first to have established actual imperial professorships, but only of rhetoric, in both Greek and Latin.[346] This reflects again the primacy of literary education in ancient culture (represented by completion of the trivium), over any kind of support for education in the quadrivium, natural philosophy, or specific sciences. The most we see of the latter are the endowments, by some emperors and municipalities, of philosophy professorships in some prestigious (or pretentious) cities, and imperial, municipal, and perhaps sometimes private support for "Museums," which were not 'museums' in the modern sense where collections are displayed, but 'Halls of the Muses', i.e. temples to the goddesses of the arts and sciences. Well before the Roman period these had become quasi-religious social clubs for scholars—although ancient Museums *did* often display statues of men renowned for their achievements in the arts, and sometimes other antiquities besides (and probably some scientific specimens and instruments), and were often associated with libraries. Most included lecture halls and communal dining rooms where 'members' ate for free. Since they were essentially the pagan equivalent of specialized churches, they had the same legal and social status as sacred ground.[347] Philostratus specifically identified a museum as a place where distinguished men who gain membership can socialize and eat free meals together, much like the upperclass 'gentlemen's clubs' of the 19th century.[348]

---

students failed to pay (a point supported by *Digest of Justinian* 50.13.1.4).

344. *Digest of Justinian* 50.5.8.4. See Trapp 2007: 19–20.

345. See Cuomo 2000: 36–37.

346. Suetonius, *Vespasian* 17–18; Cassius Dio, *Roman History* 65.12.1. Suetonius does not say how many or where, but possibly only one of each and at Rome. See discussion in Woodside 1942.

347. See *OCD* 974–75 (s.v. "museum").

348. Philostratus, *Lives of the Sophists* 1.22.524.

The most famous of these was the Museum of Alexandria. Previously established and maintained by the Ptolemaic kings of Egypt, the Alexandrian Museum received continued support from the Roman government and the city of Alexandria. Strabo says under the Romans its president was appointed by the emperor.[349] There were also Museums in other cities, such as Ephesus, Pergamum, Smyrna, Tarsus and of course Athens. Some of these appear to have had an even stronger emphasis on medical science than the Museum at Alexandria.[350] Alexandria's medical faculty had always been and remained preeminent, but its Museum must still have included in its membership engineers and astronomers, as well as scholars in other fields, as it certainly had done for centuries. The membership of the Athenian Museum, likewise, included philosophers and orators, if not other professionals as well.[351] So it would be unbelievable that Ptolemy and Hero, the greatest known writers in the physical sciences in the Roman period, would both have resided and worked in Alexandria without being associated with the Museum.

Philosophy professorships might also have been associated with the Alexandrian Museum. A certain Anatolius (whether at that time or later, a Christian) was "appointed to the school of Aristotle's successors in Alexandria by its citizens" in the third century A.D., a school that provided free meals and other "privileges" to its members (which possibly included a municipal salary or stipend in exchange for teaching). And this was the most scientific of sects. It's hard to imagine it was not connected with the Museum in some respect, as it would have been at Athens.[352] Other sects may

---

349. Strabo, *Geography* 17.1.8. On the Museum of Alexandria see Schürmann 1991: 13–32 and (though outdated in several respects) E. Parsons 1952 and Sarton 1959: 29–34, 141–57. And further notes below.

350. Nutton 1971a, 1975, and 1995; Marrou 1964: 284–91, 574 n. 15 (= Marrou 1956: 190–93, 411); and sources in von Staden 1989: 460 (esp. n. 75–76).

351. On the Athenian Museum see Oliver 1977.

352. Eusebius, *History of the Church* 7.32.6–12 and Cassius Dio, *Roman History* 78.7.3 (= *Epitome* 77.7.3). Dio reports that the emperor Caracalla disbanded the school (possibly even burned its books) and abolished its privileges in 215 A.D., though this is conspicuously not mentioned in either Herodian, *History of the Empire after Marcus* 4.8.6–4.9.9 or in the *Augustan History* = 'Aelius Spartianus', *Caracalla* 6.2, and Anatolius was appointed to it at least twenty years later, so even if the story is true (it seems unlikely), the school and its privileges must have been restored (Caracalla died in 217), as one should expect, since in every other respect

have been similarly represented. For example, an inscription of the same century as Anatolius honors a certain Flavius Dionysodorus, "member" of the Alexandrian Museum, as a "Platonic philosopher," which could mean he held the chair of its Platonism department, just as Anatolius likely did of the Aristotelian (on the same model as Athens, which I'll discuss shortly). Though from a much later date, a papyrus of the sixth century supports this inference by mentioning that a certain Asclepiades "worked all his life in the Museum" in "the great city of Alexander" as "a teacher of philosophy."[353]

Unfortunately the whole source situation for the Alexandrian Museum is surprisingly poor, especially for the Roman period, although we know from extant inscriptions and papyri that this museum and its famous library still thrived.[354] Outside Alexandria, we are told Marcus Aurelius endowed "professors at Athens in every educational field with an annual salary" in the later second century (most likely in 176 A.D.), and we know a little about what exactly this involved.[355] This endowment paid handsome salaries for

---

his purge and suppression of the Alexandrians was too extreme and universally despised to have been continued after his demise (cf. Cassius Dio, *Roman History* 78.22–23 [= *Epitome* 77.22–23] and Herodian, ibid.).

353. For Dionysodorus: Turner 1980: 86, with *Sammelbuch griechischer Urkunden aus Ägypten* 2.6012 (1915). For Asclepiades: *P. Cair. Masp.* 3.67295.

354. We do not even have the history of the institution (*On the Museum at Alexandria*) written by Aristonicus near the end of the first century B.C. (*OCD* 157, s.v. "Aristonicus (2)"), nor even *On Alexandria* by Callixeinus of Rhodes, written in the second century B.C. (*OCD* 268, s.v. "Callixeinus"); only fragments (see Christian Jacob's contribution to König, Oikonomopoulou, and Woolf 2013: 57–81). But extant papyrological evidence includes: *P. Merton* 19 (in 173 A.D. Valerius Diodorus was 'ex-vice librarian and member of the Museum'), *BGU* 3.729 and *P. Ryl.* 2.143 (144 and 38 A.D., examples of men granted the right to dine for free at the Museum for life), *P. Kron.* 4 (135 A.D. discusses certificates of membership at the library in Alexandria); see also Tod 1957: 138, Lewis 1963, and Turner 1980: 86–87 for more examples. Literature confirms these observations (see following note on the Library of Alexandria). And we have at least one inscription, declaring that in 56 A.D. Tiberius Claudius Balbillus was appointed head "of the Museum and Library of Alexandria," cf. *Forschungen in Ephesos* 3 (1912): 128.

355. Cassius Dio, *Roman History* 71.31.3: *edôke de kai pasin anthrôpois didaskalous en tais Athênais epi pasês logôn paideias misthon etêsion pherontas*. For more detail: Philostratus, *Lives of the Sophists* 2.2 (§566–67); and Lucian, *The Eunuch* 3 (here with the remark that of these "one of the two Aristotelians" had recently died, opening

a professor (or perhaps even more than one) from each of the four leading philosophical sects (Platonists, Stoics, Epicureans, and Aristotelians), as well as professors of rhetoric. Hadrian had accomplished something similar in Rome with his building and endowment of the Athenaeum. Built shortly after 134 A.D., the Athenaeum at Rome supplied professors in various subjects—though details are sketchy, it was likely similar to what Aurelius established at Athens.[356] As these setups at Athens and Rome included libraries and lecture halls as well as several chaired professorships, these really should be considered the first Western universities. A similar setup in Roman Alexandria is almost certain, especially if the Alexandrian chair of Aristotelian philosophy awarded to Anatolius was a part of the Museum and its Library, as one would expect to be the case, and we might infer the same of Dionysodorus and Asclepiades.[357] It's hard to believe Athens would have preceded Alexandria in receiving such a benefaction as endowed chairs for every major sect, so we should expect that this act by Aurelius extended to Athens a system already standardized in Alexandria.

And such a development at Athens was probably not *ex nihilo*, as all these schools with appointed professors *already* existed as private institutions that were to some extent regulated by the government. For example, in an inscription of 121 A.D. emperor Hadrian establishes, at the request of Trajan's widow (the wife of his predecessor) an allowance that non-citizens may be appointed head of the Epicurean school at Athens.[358] That the post had ever been limited to Roman citizens in the first place entails Roman state involvement from an even earlier time. The other sects likewise already had established schools and professorships in Athens as well.[359] So really Aurelius only added more centralized state organization and financing to what we should really call a loosely organized private university. The same had surely happened in Alexandria long before (in that case, before the Romans even acquired the city). Although we have no definite evidence, Athens, Rome, and Alexandria might not have been

---

up a seat, which implies there was more than one professorship for each position). See Oliver 1981 and 1970: 80–84, and Trapp 2007: 246.

356. See Boatwright 1987: 202–08.

357. Tod 1957: 138.

358. Tod 1957: 136–37.

359. Tod 1957 *passim*.

alone in receiving these benefactions and institutions.[360] But by the second century these three cities appear to have had the full practical equivalent of universities. Those early universities may have had only a limited effect on science education, as in those respective cities there were already numerous experts and philosophers to study under, and there might not have been many other cities who enjoyed such benefactions, even as a result of local or municipal philanthropy, much less imperial. But their significance should not be overlooked.

Then there are the public libraries. State and private library endowment followed similar patterns, emphasizing literary texts over scientific or technical works, and not producing much benefit to anyone who was not already well-educated and living in a major city. However, there were a *lot* of cities endowed with public libraries in the Roman period. Though already a trend among the Greeks, with many examples established by Hellenistic cities and kings, public libraries were also built or expanded in Rome itself and elsewhere by countless wealthy benefactors including nearly every Roman emperor up to the early third century, beginning with the first known public library in Rome, planned by Julius Caesar and finished by his friend Asinius Pollio in 39 B.C.[361] Besides emperors, countless other private benefactors and municipal governments followed suit all over the empire, establishing new libraries or supporting those built long before. Public libraries are known to have existed in Athens, Carthage, Corinth, Cos, Delphi, Dyrrhachium, Ephesus, Epidaurus, Halicarnassus, Knosos, Mylasa, Nîmes, Nysa, Patras, Philippi, Piraeus, Prusa, Rhodes, Side, Smyrna, Soli, Suessa Aurunca, Tarsus, Tauromenium, Teos, Thamugadi (Timgad), Tibur, Tripoli, Volsinii, and many other cities, including a public library endowed by Pliny the Younger in his native Comum, and the public library endowed by Gaius Julius Aquila

---

360. The possibility that many provinces were favored with similar set-ups is by itself plausible, but also suggested in the (albeit not always reliable) *Augustan History* (= 'Julius Capitolinus', *Life of Antoninus Pius* 11.3; and possibly implied in 'Aelius Spartianus', *Life of Hadrian* 16.8).

361. On the many libraries in Rome: Staikos 2000: 111–12. See also the recent findings from a lost work of Galen on the libraries of Rome: Nicholl 2011, Jones 2009, and Tucci 2008. In the early third century the Christian engineer Julius Africanus also "converted the Pantheon into a library for Alexander Severus" in Rome: Julius Africanus, *Kestoi* frg. 5.1 (*P.Oxy.* 3.412).

in honor of his father Celsus in Ephesus.[362] Large *private* libraries also existed in numerous locations, like those identified at Herculaneum and Pompeii, and Seneca says such collections were fashionably common.[363] Though no clear evidence of public libraries exists for Spain or Britain, and the library at Nîmes is the only documented example in France, this is most likely due to the relative sparseness of recovered inscriptions and buildings from urban areas in those regions, and the paucity of literary witnesses for them as well.[364] The remarkable frequency of libraries elsewhere argues that a public library was a common matter of pride for any major city of this period, so there must have been public libraries in many cities of Spain (certainly at the very least Cordoba and Cadiz), France (e.g. if there was one at Nîmes, then surely there had to have been one at the even more prosperous Marseilles, which was famed for its engineering schools, as noted in chapter seven), Britain (London must surely have held a public library, not least owing to the fact that the great distance of that city from any other center of learning would all but necessitate one) and elsewhere besides.

Of course the famous libraries at Alexandria and Pergamum also continued to be maintained into the Roman period. When Domitian undertook the "vast expense" to restock some libraries destroyed by fire, he "looked everywhere" for replacements, but it was to Alexandria that he

---

362. On the Comum library: Pliny the Younger, *Letters* 1.8.2, with *Corpus Inscriptionum Latinarum* 5.5262; on the Ephesus library: Hueber & Strocka 1975; on public libraries established by private benefactions in general: Platt 2008. For the other libraries named (and general discussion of Roman-period libraries) see: *OCD* 830–31 (s.v. "libraries") with König, Oikonomopoulou, and Woolf 2013; Too 2010; "Library" 2005; Staikos 2004 and 2000: 57–136; Houston 2002 and 2009; Casson 2001; Gamble 1995: 176–89, 308–17; and Fehrle 1986. For additional data (though in some cases dated): Rawson 1985: 12, 113; Wallace-Hadrill 1983: 81–82; Strocka 1981; Marrou 1964: 285 (= Marrou 1956: 188); J.W. Thompson 1962; Nilsson 1955: 49–53; E. Parsons 1952: 3–52; Götze 1937; Boyd 1915; "Bibliotheken" 1897; and Pliny the Elder, *Natural History* 35.2.10. The general destruction has made it difficult to be certain but there is also evidence of a public library at Pompeii (Richardson 1977). On the architecture of ancient libraries in this period: Johnson 1984 and Makowiecka 1978.

363. Seneca, *On Tranquility* 9.5. See Marshall 1976 for an extended discussion of the creation and use of private libraries in the Roman Empire. And Houston 2009 for examples attested in the papyrological record.

364. For example, Hanson 1989.

sent men "to copy and correct" them, demonstrating that the Alexandrian libraries were not only intact, but still had no peer.[365] Claudius might even have expanded them.[366] Pergamum already had a renowned library, but Hadrian added another in the Temple of Asclepius there, which might indicate a specifically scientific focus (if this association with Asclepius implies an emphasis on medical literature). Hadrian also established a new public library and Museum in Athens in 131 A.D., creating (as I noted earlier) what is essentially a university whose facilities were further supported or expanded by Antoninus Pius and Marcus Aurelius.[367] A fragmentary inscription mentions another public library at Athens in the Ptolemeion, where the students of a local city school had to produce 100 new scrolls every year for its collection.[368] There are numerous other examples that could be given. It is reasonable to assume that most if not all major cities had at least one public library established and maintained at state or private expense.

It was clearly a mark of prestige for a city to have a public library, and for a state or private benefactor to bestow one. Libraries represented not only an enormous initial expense, they also required considerable expense to maintain, and thus their frequency in the Roman world confirms that elite social values fully embraced the widespread preservation of and access

---

365. Suetonius, *Domitian* 20. For scholarship on the Alexandrian libraries: Nesselrath 2012; Staikos 2004: 1.157–248 and 1.283–88 (with McKenzie 2007: 50); Chapman 2001; Staikos 2000: 57–90; MacLeod 2000; El-Abbadi 1992 (with El-Abbadi & Fathallah 2008); Blum 1991; Canfora 1987; E. Parsons 1952. See previous note for mentions in Roman-period papyri. Mentions in Roman-period literature include: Strabo, *Geography* 13.1.54, 17.1.8; Galen, *Commentary on the 'Epidemics' of Hippocrates* 3; Philostratus, *Lives of the Sophists* 22.3, 25.3; Athenaeus, *The Dinnersages* 15.677e; and the *Augustan History* = 'Aelius Spartianus', *Life of Hadrian* 20.2.

366. Suetonius, *Claudius* 42 ("adding to the Museum" in the middle of a fully-developed city seems an insignificant honor unless this meant an expansion of the Museum's library; that the Library was located within the Museum is confirmed by Athenaeus, *The Dinnersages* 1.3a; and implied by the content of Strabo, *Geography* 17.793–4).

367. Pausanias, *Description of Greece* 1.18.9. See Oliver 1977 (esp. p. 166 n. 10) and Boatwright 2000: 153–57 and Staikos 2000: 125.

368. This inscription even includes part of the catalogue of the library's collection: Gamble 1995: 182 and Marrou 1964: 285 (= Marrou 1956: 188).

to the written word. Though public libraries were primarily stocked with literary works outside the field of science and philosophy, the latter would still have been adequately represented in most collections, particularly in cities with a strong scientific presence, as it would have been a matter of pride (and self-promotion) for scientific authors to gift their local libraries with editions of their latest works. So the presence, prestige, and support of public libraries does indicate some support for the preservation and availability of works in science and natural philosophy, which provided an avenue for some laypersons to expand their education in the sciences. However, library books were only of use to the very small segment of the population that was literate, and not merely literate, but competent and educated enough to read, comprehend, and make use of advanced works in philosophy and science. Moreover, public libraries did not lend out their books. Patrons had to read books in the library and leave them there. This meant access to a private patron's library would still be more valuable than living near a public library, since borrowing books from a friend or patron could be arranged, likewise staying on as their guest.[369] Nevertheless, the widespread existence of public libraries is again something remarkably unique, and unparalleled in any other ancient culture. And like the comparably unusual phenomenon of regular public orations on all topics, the effect of public libraries on disseminating scientific knowledge would have been similar—which means similarly limited, but still notably greater than in any other ancient culture we know.

Finally, there were a scant few attempts to educate the public through educational inscriptions, and though these could only be read by the literate, those who did read them would spread their news further by word of mouth. Even so (except for the equivalent of weather almanacs), such inscriptions were too rare to have any general impact.[370] The emperor Claudius, for example, attempted, as Cicero would have put it, to "banish

---

369. See discussion in Marshall 1976.

370. Ancient inscriptions serving as practical weather almanacs were ubiquitous (as were wind-roses and sundials). Called *parapêgmata*, these almanacs combined a calendar with related astronomical data and meteorological expectations. See Taub 2003: 20–37, 41–43, 173–76; and Lehoux 2007. Some cities also had public mechanical clocks, often with astronomical and calendric features, e.g. Schürmann 1991: 258–72 and Noble & de Solla Price 1968, whose mere existence may have disseminated some scientific knowledge.

fear and false religion from confused men" with a science lesson, by erecting an elaborate public inscription in the early first century A.D. explaining the scientific causes of an approaching solar eclipse predicted by astronomers.[371] He hoped this would forestall superstitious reactions to the event that might threaten public order. Though nothing alarming did happen, there is little indication his science lesson seeped very far into public consciousness. In the same fashion, there would have been a lot of rather sensible natural philosophy in the monumental inscription at Oenoanda commissioned by Diogenes, all on his own dime, spelling out the entire philosophical system of Epicurus. But this was just as rare and had no more apparent impact on public consciousness.[372] Nevertheless, inscriptions like these indicate an occasional value for public science education well beyond the norm.[373]

All in all, state and public support for education in the early Roman period was not as extensive as that seen in nations after the Scientific Revolution, but in many respects greater than any seen throughout most of the Middle Ages or in any other culture (other than the Hellenistic culture

371. Cassius Dio, *Roman History* 60.26.

372. Diogenes of Oenoanda, *Epicurean Inscription* = M.F. Smith 1996 (cf. *OCD* 457, s.v. "Diogenes (5)," and *EANS* 253–54, s.v. "Diogenes of Oinoanda"). See also Warren 2009: 54–59.

373. A related example, though a century before our period of interest, was the tomb of Archimedes, which at his bequest was inscribed with one of his mathematical theorems—the inscription was subsequently lost in Sicily until rediscovered and restored by Cicero: Cicero, *Tusculan Disputations* 5.23.64–66 (cf. Plutarch, *Marcellus* 17; Simms 1990; Cuomo 2001: 197–98; Jaeger 2002: 55–56). Around the same time Eratosthenes dedicated an inscription in Alexandria popularizing his scientific invention of the mesolabe (a kind of slide-rule for calculating scaling functions for architects and engineers: see Russo 2003: 111; Netz 2002: 213–15; Knorr 1989: 131–53; Cohen & Drabkin 1948: 62–66). We also know of two inscriptions from the Roman period that publicized astronomical theories, though in an advanced form that would have been unintelligible to laypeople: the so-called "Canobic Inscription" erected in 146 A.D. (containing "the first principles and models of astronomy" dedicated "to the Savior God," as many deities were then named, by "Claudius Ptolemy," the famous astronomer himself), of which we only have a textual transcription, and the "Keskinto Inscription" (*Inscriptiones Graecae* 12.1 §913) erected in Rhodes around 100 B.C. On these see Jones 2006, Evans 1999: 384–85, and Hamilton, Swerdlow, and Toomer 1987. For another inscription of the Roman period praising geometrical and astronomical science, which may be linked to the family of Galen, see Nutton 2004: 216–17.

that originally influenced the Romans, and perhaps the Byzantine culture after them). This included the first steps toward what were effectively universities (with all the components in place, just lacking, perhaps, centralized administrations), and a widespread commitment to making the written word available in diverse subjects through public libraries, which would have included works in the sciences. Likewise, although state-supported education did not aim at improving access to science education, it would have indirectly done so to some degree, especially in state support of Museums, which provided more direct support for associations of scholars and scientists, and in the endowment of professorships in philosophy and medicine, and the extending of privileges to them. Medieval state and public support for education is not likely to compare as well, until the rise of the universities, yet even those were small and few in number for quite some time and thus, at least until the Renaissance, might not have surpassed what had already been available in the early Roman empire.

9. Jewish and Christian Education

So far as we are able to tell, the availability of education among the Jews of the Roman Empire was not significantly different from that among their Greek and Latin neighbors. The elite were privileged, the poor were left out. As Lapin observes, "rabbis, at least those about whom stories are told in the literary sources, appear to have been wealthy, and the most prominent centers of rabbinic activity appear to have been Palestinian cities."[374] In contrast, the first Christians supposedly came from rural districts, and even when the movement went urban its base of support remained among the poor or working class. The very leaders of the movement, Peter and John, as portrayed in the Gospels at least, were not atypical Jews of their day (in comparison with the Palestinian population as a whole), and yet we are told they were "illiterate laymen."[375] That is probably fiction—they were more likely highly educated Rabbis.[376] Similarly, the Gospel of John shows a Jewish audience reacting with surprise at the fact that Jesus can read.[377] Also

374. Lapin 1996: 505 (cf. 505–08).

375. Acts 4:1–6. The word *agrammatoi* literally means "without letters," hence unable to read or write (*LSG* 14, s.v. "agrammatos" I), while *idiôtai* means without professional training or knowledge (see note in chapter two).

376. Carrier 2014: 263–64, 440.

377. John 7:14–18. The mention of Jesus drawing something on the ground in John 8:6–8 is ambiguous and generally regarded as not even original to the Gospel of John (called the *pericope adulterae*, it has been identified as a later interpolation). It cannot be known on present evidence if Jesus, granting that he was historical

probably fiction. But even then, both tales were meant to reflect a wider reality. The gulf between the elite and non-elite in Jewish society was as wide as in Roman society generally.

While among the Jewish *elite* a teacher of the Bible held a social status comparable to the philosopher among the Greco-Roman elite, the *common* people could actually regard the bible scholar with immense hatred.[378] According to the Talmud, "The rabbis taught that a man should sell all his possessions and marry the daughter of a scholar," for then his children can become scholars, too, but a man must not marry the daughter of a "commoner," because commoners are no better than beasts, and their children will become commoners instead. In this context, Rabbi Akiba recalled that "when I was a commoner I said, 'If I could lay my hands on a scholar, I would maul him like an ass!'" and Rabbi Eliezer said "if the commoners did not require us for their own welfare, they would kill us!" Likewise, Rabbi Hyya taught that "a man who occupies himself with the study of the Law in the presence of a commoner evokes as much hatred from him as if he had stolen his bride," for "the enmity of a commoner toward a scholar is even more intense than that of the heathens towards the Israelites, and the hatred of their wives even greater than that!"[379]

This implies a very stark divide indeed. There is no good evidence that Jewish literacy was any greater than Greek or Roman literacy, and we have already seen how limited that was. Some scholars have attempted to push the case for greater, even universal literacy among the ancient Jews, but their evidence does not hold up, a fact that has been thoroughly demonstrated by Catherine Hezser.[380] Alan Millard is among those claiming more

at all, could read, since the authors of the Gospels, believing Jesus was divine, might simply assume a god could read. But if Jesus was an actual Rabbi (Mt. 26:25, 26:49; Mk. 9:5, 11:21, 14:45; Jn. 1:38, 1:49, 3:2, 4:31, 6:25, 9:2, 11:8), he would almost certainly have been literate—and therefore probably (in reality) from a family of some means, regardless what the Gospels claim.

378. On the higher status of Bible scholars among the elite: Marrou 1964: 454–55, 616 notes 6–7 (= Marrou 1956: 316–17, 445). On debates surrounding the meaning and provenance of the following passages see Rubenstein 2003, pp. 200–201, esp. n. 68.

379. All from b.Talmud, *Pesachim* 49a-49b.

380. Hezser 2001. See Quick 2014 for a survey of subsequent scholarship on Jewish education and literacy, especially Rollston 2010, though none of it challenges

widespread literacy.[381] But his methodology is flawed in obvious ways that will be apparent to anyone who looks at his evidence in light of the sounder analysis of the same kinds of evidence elsewhere.[382] In the end, all of his evidence either does not support his conclusion or only proves literacy among the elite and well-to-do, exactly what we find in the Roman context. And Millard's case, even at its best, only supports Hebrew or Aramaic literacy, with almost exclusive focus on scripture, and as far as we know very little of any science or natural philosophy was written in Hebrew or Aramaic, and none of any significance in scripture. Like Millard, Gamble also claims widespread literacy among the Jews, but every passage he cites as evidence for this actually fails to say what he claims, or once again only supports literacy among the elite and well-to-do, or even confirms the reverse, establishing that Jewish children were taught scripture and religious law *orally*, not by reading. Indeed, Gamble otherwise concludes that, starting in the second century, the Christians put their new adherents through a system of "doctrinal and moral instruction" that "certainly did not include learning to read or write" but "did include close familiarization with Christian scripture," and these "catechetical schools" may well have been modeled on their Jewish counterparts.[383]

The only significant evidence for wider Jewish education is a single passage in the Talmud reporting that in the early 60s A.D. (and apparently continued for centuries thereafter) certain Jewish authorities started subsidizing education, first in Jerusalem, and then in every Palestinian city (and possibly in a few other places), although the same passage reports difficulties in getting students to attend.[384] This looks like an emulation of the Greco-Roman educational foundations that we already examined in the previous chapter. So these Palestinian schools were probably, like their Greek and Roman equivalents, not really available to everyone or, as the references

---

Hezser.

381. Millard 2003a and 2003b (with Carrier 2003) and Millard 2000: 154–84.

382. See Harris 1989: 281–82, Marrou 1964: 373–75 (= Marrou 1956: 254–56), and most decisively Hezser 2001: 39–109.

383. Gamble 1995: 6–8. An even bolder case is made in Safrai 1969, but Safrai's account of Jewish education is wholly unreliable and his discussion of the evidence often wildly inaccurate.

384. b.Talmud, *Baba Bathra* 20b-21a.

to hostile "commoners" above suggests, not widely attended. The numbers given in the Talmud imply only a few dozen boys, at most, attended in any given city. Moreover, literacy is not mentioned. The subsidized schools in question only taught the Torah, and thus could have been schools of oral instruction in scripture and religious law, as later Christian catechetical schools were. In fact, oral instruction is all the Torah actually commands, a commandment in effect for centuries before any kind of formal schools were proposed, which is why most religious education must always have taken place in the home.[385] In fact, most of the Jewish law was contained in the Mishnah (the "Oral Torah"), which was not even written down until around 200 A.D. and thus until then certainly wasn't taught in any other way *but* orally. And the Talmuds, which are commentaries on the Mishnah, were not written down for another century or more after that. The scripture itself was regularly read out to audiences at synagogues and therefore learning it did not require literacy.[386] Thus "schools for religious instruction" would not even imply the teaching of literacy.

Regardless of which schools are meant, the content of all Jewish schools, even at the most advanced levels, was little more than scripture and Jewish law, and in Palestine limited only to Hebrew or Aramaic.[387] This meant any special Jewish education of the time included nothing of science or natural philosophy but was entirely consumed with the study of scripture and law, and the interpretation thereof, and would not equip the student to study science at all, as that would generally be taught, in lectures and speeches and books, only in Greek. Although "biblical interpretation" (like commentaries on the poets in pagan schools) might have occasionally touched on science content (such as when debating the age of sexual consent for young girls[388]), Jews were actually admonished by their own authorities with the proverb "cursed be a man who teaches his son Greek wisdom." It was allowed that

385. Deuteronomy 6:1–2, 6:7, 6:20–25.

386. See Hezser 2001: 94–109 (with *OCD* 1419, s.v. "synagogue").

387. Lapin 1996: 498–511; Gerhardsson 1961: 56–66 and 85–92; and conceded even by Millard 2000: 158.

388. See Meacham 2000, although reliance on actual science in this debate was minimal at best. By contrast see my related discussion in chapter four of science content in Roman jury trials; and, in chapter six, of science content in Roman law schools.

"the Greek language and Greek wisdom are distinct," and "they permitted the household of Rabbi Gamaliel to study Greek wisdom because they had close associations with the Government," but the fact that these kinds of distinctions were being made entails that in most cases studying science or natural philosophy was either frowned upon or not encouraged, and at best was truly exceptional when it occurred.[389] In consequence, as one might expect, the Talmuds evince a love-hate relationship even with scientific medicine.[390] Indeed, there are no clear cases on record of Jewish research scientists in the Greco-Roman sense (although they must have had at least some practicing scientists in all active fields).[391]

But that only represented the Palestinian and/or the most traditional Jewish point of view. In the Diaspora, by contrast, Greek education, where available, had been widely embraced by Jewish communities, to the point that there may have been little difference between educated Greeks and Jews in terms of what they studied or learned.[392] If Philo of Alexandria is at all representative of an elite Alexandrian Jew in the first century, their views on education were not so different from the Greeks or Romans, except for being a lot more piously focused on scripture and mystical wisdom than anything like empirical science. Philo clearly had attended Greek schools and completed the entire *enkyklios* and went on to study philosophy and fully approved of having done so. That was probably no more common or unusual for an elite Diaspora Jew than for an elite Alexandrian pagan. We have already noted Philo's views on this in chapter five, but we can add more here.[393]

Philo argued that scientists and natural philosophers ought to dedicate their study of nature to God, just as a sailor dedicates the success of a voyage to God, showing humble thanks for the talent God had given them, which God obviously intended them to use. This means Philo had no problem with Jews becoming scientists and studying science, so long as they honored

---

389. b.Talmud, *Sotah* 49b. For more discussion see sources cited in Judge 1983: 9.

390. See Newmyer 1996 (and following note).

391. There were some Jewish scientific doctors: see Kudlien 1985 and Rosner 1994.

392. See Gruen 1998 and 2002.

393. For a detailed treatment of Philo's views on education, see Sandnes 2009: 68–78.

God as they ought.[394] This exemplifies a pious attitude that did not require the depths of hostility that many Christians reached (as we shall soon see) and that many Palestinian Jews apparently shared (as we just saw).

Instead, Philo embraced an honored place for men of science. In his scheme of values:

> Some men are born of the earth, and some are born of heaven, and some are born of God. Those born of the earth are hunters after the pleasures of the body. ... But those who are born of heaven are men of skill and science and devoted to learning. For the heavenly portion of us is our mind, and the mind of every one of those persons who are born of heaven studies the encyclical branches of education and every other art of every description, sharpening, and exercising, and practicing itself, and rendering itself acute in all those matters which are the objects of intellect.[395]

However, those born of *God* are even more superior still—they are the priests and prophets who actually *abandon* intellectual studies for the loftier revelations of God. He gives the specific (though apocryphal) example of Abraham, who supposedly devoted his life to the study of astronomy and cosmology before he was called by God, and was thus a man of heaven, but once he was called, he abandoned these pursuits to become a man of God.[396] Philo clearly describes this as an improvement, just as he refers to abandoning scientific and philosophical studies for, instead, a life of pleasure and vanity, as abandoning the better for the worse.[397] Thus, though Philo imagined an elevated place for science and awarded it respect and acceptance, he still thought that, ultimately, one ought to abandon science and prefer the contemplation of God—specifically declaring, for example, that continuing to study astronomy is only second best to fully embracing a life of God.[398] Philo even denounces theoretical astronomy as a waste of time (in terms very similar to those Christians would later amplify) in a rather long rant

394. Philo of Alexandria, *On the Change of Names* 39.219–22.

395. Philo, *On the Giants* 13.60.

396. Philo, *On the Giants* 13.61–14.63. See also Philo, *On the Change of Names* 9.66–68 & 10.76.

397. Philo, *On the Giants* 15.65. A similar staged scheme of values is described in Philo, *Who is the Heir of Things Divine* 9.45–48, 20.96–99, 22.108–23.116.

398. Philo, *On Mating with the Preliminary Studies* 9.47–49, 10.51.

concluding, "Oh fellow, do not think about things beyond, but attend only to what is near you, and examine only yourself instead," so that in fact "if you really want to be a philosopher," just study the sciences of man, learning the nature of your body and mind—he even specifically cites the example of Socrates, who abandoned natural science for self-contemplation.[399] Apart from this slackening of favor for scientific knowledge, Philo and no doubt most elite Diaspora Jews *almost* embraced the same educational values as most Romans of comparable social station, at least with respect to the ideal place of science and natural philosophy. This stands in significant contrast to the attitudes of Christians in the same period.

Christianity was an offshoot of Judaism, in many respects a syncretism of Jewish and pagan beliefs (particularly after the first century).[400] Most educated Christians before the fourth century were converted after receiving a pagan (or sometimes Jewish) education, and given the emphasis of evangelists on recruiting women, and men among the working classes, most converts will have received no education at all, or very little, and most of those who did will only have received a primary or secondary education, hence very few would have completed any higher education before conversion. This would still be true even if there was not an evangelical emphasis on women and the working class (although there appears to have been). As shown in chapter two, Christian intellectuals of the third century believed their converts represented a cross-section of the whole society, and just as most Romans were illiterate, and most of those who were literate had not completed enough education to gain much exposure to science or natural philosophy, it follows that most Christians were illiterate and most literate Christians did not have much exposure to science or natural philosophy. Indeed, even the skills and practice of logic and creative or critical thinking were only significantly taught in higher education, which few Romans, hence few Christians, ever attained.[401]

---

399. See Philo, *On Dreams* 1.10.(52–60).

400. See Klauck 2003; Carrier 2011 and 2014; Fox 1987.

401. Corroborated by Sandnes 2009: 5–7. One might hear the occasional crank claim that 1 Corinthians 4:15 supports a broad base of education among Christians, but that passage says only "if you *were* to have tens of thousands of tutors in Christ," which is a subjunctive counterfactual, meaning they did *not* have tens of thousands of tutors. Moreover, the word "myriad" often simply meant "countless"

But we can still ask how much and what kind of education predominated among those Christians who came to the faith before receiving any education in science or reasoning. Did Christianity encourage education of any sort, or discourage education in any way? Surveying Christian attitudes toward basic education in the early Roman empire, Harry Gamble found that even rudimentary education of the day "was predicated on pagan texts infused with moral and religious ideas of which Christians disapproved, and this discouraged some who might otherwise have taken advantage of it." In fact, "well-educated Christians usually received their education before conversion and were themselves unable to recommend pagan schooling to those who were already Christian." Worse, "many Christians were suspicious of 'the wisdom of this world' [following 1 Corinthians 1:20–21, 2:6, 3:19], and among them there was a tendency" either to "sanctify ignorance" or at least "neglect education in the interest of fideism, otherworldliness, or acquiescent orthodoxy."[402]

I will discuss Christian attitudes toward science more broadly in *The Scientist in the Early Roman Empire*. Here my focus is on education. Even though Christian intellectuals like Tertullian could just barely rationalize Christians becoming students, he still forbade Christians from teaching, because the educational process was too pagan.[403] He does not even think of reinventing education along a Christian model. Hippolytus was only slightly more lenient, since he was willing to let it slide if a Christian teacher had no other way to make a living, but otherwise he said it was immoral to be a teacher.[404] This sort of hostility declined over subsequent centuries.[405] But

---

and is thus a hyperbolic expression, while "tutor" here is not "teacher" but "nanny," literally the *paidagôgos*, a slave that took well-to-do children to school (*LSG* 1286, s.v. "paidagôgos"), which Paul elsewhere uses as a metaphor for the Jewish law (Galatians 3:24–25), rather than an actual tutor. For a discussion of this and other education metaphors in the New Testament (and an important analysis of Paul's latent hostility to higher education) see Judge 1983.

402. Gamble 1995: 6. See also Jacobs 2011 and Sandnes 2009.

403. Tertullian, *On Idolatry* 10. See Sandnes 2009: 111–23.

404. Hippolytus, *Apostolic Tradition* 2.16.5.

405. As summarized in Jacobs 2011. And besides following notes, see Pailler & Payen 2004: 265–67 and Sandnes 2009 (and, tangentially, Hauge & Pitts 2016 and Dutch 2005). Note also that Christians remained canonically opposed to allowing *women* to teach or lecture (on any subject, much less religion), following

a strain of hostility remained influential. Hence a third century Christian tract says:

> Avoid all books of the heathen. For what have you to do with strange sayings or laws or lying prophecies, which also turn away from the faith those who are young? For what is wanting for you in the word of God, that you should cast yourself upon these fables of the heathen? If you want to read historical narratives, you have the Book of Kings. If wise men and philosophers, you have the Prophets, wherein you shall find more wisdom and understanding than of [the so-called] wise men and philosophers, for they are the words of the one God, the only wise being. And if you wish for songs, you have the Psalms of David. But if you'd rather read about the beginning of the world, you have the Genesis of the great Moses. And if you want laws and commandments, you have the glorious Law of the Lord God. All that is strange therefore, everything contrary [to the Bible], wholly avoid.[406]

This was not an unusual sentiment. Reading or learning about science and natural philosophy was useless or even harmful, since whatever you need to know is already in the Bible, the only moral and reliable book there is.

One might expect this would lead to a new educational curriculum based on the Bible. Yet in the early Roman empire, Gamble observes, "the ancient church never undertook an alternative system of education for the faithful."[407] Ellspermann came to the same conclusion, finding that "the Church did not attempt in the first three centuries to establish schools for children," or even propose or discuss the idea.[408] There were a few distinctly

---

1 Corinthians 14.34–35 and 1 Timothy 2.11–12, on which see Levick 2002: 151–53. And adding to Jacobs, see Too 2001: 405–32 for Sara Rappe's analysis of the Christian struggle to assimilate pagan education during the transition to the Middle Ages, while Kaster 1988: 70–95 discusses the divergence on this point between Eastern and Western branches of medieval Christianity.

406. *The Catholic Teaching of the Twelve Apostles* 1.6, here modernizing an older English translation (from the extant Syriac translation of the Greek original, which may date back as far as the second century) in Connolly 1929: 12; cf. *ODCC* 479 (s.v. "Didascalia Apostolorum"). See Sandnes 2009: 102–10.

407. Gamble 1995: 6.

408. Ellspermann 1949: 1–3. For a survey of early Christian education see Marrou 1965: 451–71 (= Marrou 1956: 314–29).

"Christian" schools of higher education by the third century, at least in the sense of studying under a Christian intellectual like Origen (which I'll soon discuss). And there were Christian "catechetical" schools, which were also for adult converts, not children, but these assumed a position more similar to Jewish schools for training in religious principles and ideology (which I discussed previously). These do not appear to have involved or required even literacy or any other prior schooling, but consisted only of oral instruction in religious dogma.[409]

As a result, Gamble observes, "Christian writers of the first five centuries acknowledge a standing distinction between a small number of literate and intellectually active Christians and the majority of believers" and "all indications are that the extent of literacy within the church did not exceed its extent in society at large, and during the first three centuries it probably did not attain even that level."[410] As Paul observed of the early Church, "not many who are wise according to the flesh, not many powerful, not many wellborn, are called" to the faith.[411] In fact, the hostile Celsus claimed Christians went around telling students that schools only taught "empty nonsense" and "nothing good" and "if they want, they ought to send their fathers and teachers away" so they can "attain perfection."[412] Origen responded to this charge by admitting that Christians really do "turn people away from teachers who teach iambic verse and whatever else that neither corrects the speaker nor benefits the listeners," but they do not turn people away from virtuous teachers or classes in philosophy, except by insisting that their own teaching is better, which is a tacit admission of the very thing Celsus was talking about.[413] Literacy could not be acquired in the ancient educational system without studying verse, as that was the only method used to teach it. And without literacy, higher forms of education were

409. See Pack 1989 and Clarke 1971: 122–23.

410. Gamble 1995: 6–7. His conclusions are corroborated in Sandnes 2009.

411. I Corinthians 1:25–30.

412. Origen, *Against Celsus* 3.55. Note that this Celsus is likely the Epicurean friend of Lucian (addressed in Lucian, *Alexander the Quack Prophet* 1–3 and 60–61; cf. Origen, *Against Celsus* 3.35), but probably not the same as the engineer of the same name and similar date, and certainly not the same as the encyclopedist Aulus Cornelius Celsus, who dates a century earlier.

413. Origen, *Against Celsus* 3.56–58.

simply unattainable. The reality was that most Christians did not see very much need for education, since salvation could be had without it, and what else mattered?

At the end of our period of interest (the dawn of the fourth century), Eusebius wrote that pagans "go up and down and all around constantly talking about education, saying above all that those who are going to make any attempt to grasp the truth must pursue astronomy, arithmetic, geometry, and music," in other words, the quadrivium, the only course of secondary education with any significant science content. As Eusebius complains, pagans assert that "lacking in these a man will not be able to perfect himself as a learned philosopher," in fact, "he will not even be able to touch the truth of things, unless the knowledge of these subjects has first been impressed upon his soul," echoing exactly what Strabo had said (as we saw in chapter five), and reflecting the sentiments we have seen from every elite author we have examined. Instead of agreeing, Eusebius mocks these educational values:

> Those who stretch out and hold up this education of which I speak, they think they are lifted up high in the sky to walk on the very ether, as if carrying God himself in their numbers, while they conclude that we, because we do not pursue the same studies, are no better than cattle, and say we cannot know in this way either God or anything majestic.[414]

Ironically, this is almost the exact sentiment voiced by the Roman astronomer Ptolemy *in praise* of science education, in his famous epigram, "I know I am mortal and fleeting, but when I search the densely revolving spirals of the stars, I no longer touch the earth with my feet, but dine with Zeus himself, and take my fill of ambrosia," the food of the gods (which conferred immortality on those who ate it), and yet here this sentiment is *mocked* by a Christian instead of being emulated or admired.[415] Eusebius thus voices a typically anti-elitist sentiment, positioning himself with the uneducated Christians who are insulted by the claim that they cannot understand the truth without an education—even though that happens largely to be true.

414. Eusebius, *Preparation for the Gospel* 14.10.10.

415. Ptolemy's epigram survives in the *Palatine Anthology* 9.577. Galen said essentially the same thing of a medical education, e.g. Galen, *On the Uses of the Parts* 3.10 and 17.3 (= May 1968: 189–91, 733).

And though Eusebius certainly had a far superior education than any of the Christians he is siding with, he is nevertheless sneering at the very idea that anyone needs such an education. This flips quite upside down the attitude to education we have seen from the pagan elite. And this was not atypical within early Christianity.[416] The very arguments pagans used in *favor* of education, Eusebius sees as an insult to Christians (who have no need of an education), rather than as a reason to promote a broader access to education.

In a similar fashion Lactantius, one of the most famous Christian educators in the first three centuries, and eventual tutor to the imperial family itself (he advised Constantine, who would become the first Christian emperor in 313 A.D., and tutored his son), said philosophy cannot be wisdom *because* only the very learned can understand it, and surely God would ensure that wisdom was common and easily available to all.[417] Therefore, he argues, only the Gospel is wisdom, and therefore the Gospel is the only thing all people should be taught. And receiving the Gospel did not require any education in science or philosophy, or even literacy.[418] So it would certainly seem Celsus was accurately describing a prevailing Christian attitude toward education.

We even have a first-hand account of this attitude. When Justin Martyr went out to study philosophy, obviously having completed primary and secondary school as a pagan (but not the *enkyklios*), he toured the sects and did not like them, eventually falling into Platonism because it was close to what he was looking for, but ultimately converting to Christianity and giving everything else up.[419] Justin's account is not entirely believable, but even as fiction it reflects his educational values. First Justin took up with a "very celebrated" Pythagorean. This already sounds like fiction, since there was barely any such thing at the time, and surely had Justin studied under a famous man, he would have named him. But whether truth or fiction, as Justin tells it:

> When I had an interview with him, willing to become his hearer and disciple, he said, "What then? Are you acquainted with music, astronomy,

416. See the analysis of Copan 1998.

417. Lactantius, *Divine Institutes* 3.25.

418. Lactantius, *Divine Institutes* 3.26–27.

419. Georges 2012, Ulrich 2012, and Sandnes 2009: 84–95.

> and geometry? Do you expect to perceive any of those things which conduce to a happy life, if you have not been first informed on those points which wean the soul from sensible objects, and render it fitted for objects which appertain to the mind, so that it can contemplate that which is honorable in its essence and that which is good in its essence?" Having commended many of these branches of learning, and telling me that they were necessary, he dismissed me when I confessed to him my ignorance. Accordingly I took it rather impatiently, as was to be expected when I failed in my hope, the more so because I deemed the man had some knowledge. But reflecting again on the space of time during which I would have to linger over those branches of learning, I was not able to endure longer procrastination.[420]

This is exactly what Eusebius and Lactantius refer to: many philosophers expected students to study the sciences, at least the quadrivium, before they could understand any more complex subjects. Instead of seeing the point, and rolling up his sleeves and getting to the hard work of learning something, Justin simply leaves and looks for an 'easy' school instead.

Justin had previously opted not to study Aristotelianism, because, he claims, Aristotelians required a fee—though this would have been no more or less true of any other school, and in fact the Aristotelians would have required completion of the *enkyklios* as well. But clearly, Justin wanted a school that required him to do no work, and whose teachers required no pay. And he rejected Stoicism because, he claims, Stoics were agnostics and dismissed theology as a subject—though in fact the Stoics were not commonly agnostics and promoted extensive views on the subject of theology, they *were* too scientific for a Christian (as we'll see when we hear from Origen).[421] He eventually settles in with a Platonist, even though any serious Platonist would have set the same preparatory science requirements

420. Justin Martyr, *Dialogue of Justin and Trypho the Jew* 2. Note the "Pythagorean's" reason for requiring the *enkyklios* is precisely that it increases Hypothetical-Categorical-Abstract Reasoning (see note in chapter five), i.e. the ability to think abstractly and in hypotheticals, essential to science.

421. In fact the earliest New Testament theology was very akin to the Stoic: M. Lee 2006; Engberg-Pedersen 2000 and 2010 (on which see the critical exchange in *Journal for the Study of the New Testament* 33.4 [2011]:406–43), and Rasimus, Engberg-Pedersen, and Dunderberg 2010. But Stoic theology emphasized empiricism over dogmatism, and by being based on contemporary physics, was too materialistic and unspiritual for the likes of a mystic such as Justin.

that Justin attributes to the Pythagorean. So his entire account is absurdly fictionalized. But this still reveals that Justin had no interest in any education that required time or money—in other words, he had no interest in any real education. He wanted quick and cheap 'wisdom', and in the end that's what he got from the Christians. This very definitely reverses the educational values of the pagan elite. Attitudes like this were sure to lead more Christians away from scientific content than toward it, and dramatically more so than for anyone else in the ancient educational system.

There were still exceptions, and thus early Christianity managed to secure a few educated intellectuals who did not spurn the idea of reading a book other than the Bible. But many Christians who took any more positive interest in learning science or natural philosophy were actually branded heretics or the associates of heretics—moreover, there were not very many of them. For example, Paul of Samosata attacked a small enclave of Christian "heretics" because they greatly admired and devoted themselves to the study of the scientists Euclid, Aristotle, Theophrastus, and Galen—he even blamed their heresy on this very devotion.[422] I'll discuss more instances of this attitude in *The Scientist in the Early Roman Empire*, but the most notable example is what happened to the Christian school of higher education established by Origen—who was later declared a heretic.[423] This was set up in Caesarea in the early third century, somewhat in opposition to a more orthodox form of Christian higher education offered at Alexandria, though both of these were highly modeled on the schools of Platonists.[424] These are the only two distinctly Christian educational institutions we know of before the fourth century.[425]

---

422. Eusebius, *History of the Church* 5.28. See discussion in Walzer 1949: 75–86 and for background see *ODCC* 1242 (s.v. "Paul of Samosata").

423. For background on Origen see *OCD* 1047–48 (s.v. "Origen (1) (Origenes Adamantius)") and *ODCC* 1193–95 (s.v. "Origen" and "Origenism"). See also Jacobs 2011 for Christian debate over the value of Origen's educational ideals in the 4th century; and Gemeinhardt 2012 on the Origen-Gregory correspondence and its relation to evolving educational values in Christianity.

424. Little is known of the Alexandrian curriculum, but for discussion of what might have went on at that Christian school at Alexandria see van den Broek 1995 and Osborn 2005: 19–24.

425. The attitudes of these two schools to education (judging from the works of Origen of Caesarea and Clement of Alexandria, respectively) are well analyzed in

We are fortunate to have a description of Origen's curriculum from his more acceptably 'orthodox' student Gregory, in a panegyric delivered sometime in the middle of the third century. According to Gregory, Origen positioned philosophy as only a preparation for the *real* goal of advanced studies, which was scripture. Gregory had already received a full course of primary and secondary education, and had even begun studying the advanced fields of rhetoric and law before becoming Origen's student.[426] As Lapin observes, Origen confirms the primacy of scriptural studies in a letter to a pupil, probably this same Gregory, where "Origen gives the study of scripture a rather stronger emphasis than did Gregory" in his oration, "even as Origen acknowledges the importance of the curriculum of 'preparatory' studies."[427] But Gregory praises the whole curriculum, which as Lapin says, "begins with argument (*dialectic*), then moves on to physical studies, geometry, and astronomy, as well as practical ethics" and then "theology" and finally "only then does the curriculum turn to scripture."[428] Notably, this is exactly the kind of demanding curriculum Justin and other Christians rejected.

In Origen's own words:

> Your talent could make you a perfect Roman lawyer or a Greek philosopher of the usual reputable schools, but I wanted you to exhaust the whole power of your talent on Christianity as your end, and because of this I prayed successfully that you would take from Greek philosophy those things that would enable you for Christianity, in a sense, as a general or preparatory curriculum, and also those things from geometry and astronomy that are useful for interpreting the holy scriptures, so what the

Sandnes 2009: 124–59.

426. Gregory Thaumaturgus, *Panegyric Oration on Origen* 1 and 5. There is some dispute as to the actual identity of this author, but his identification with Gregory the Thaumaturge is supported by Eusebius (*History of the Church* 6.30), who was using Origen's library at the time (Carriker 2003) and thus would be in a good position to know, while arguments against the attribution are not very persuasive. Whatever his name, the author was certainly a student of Origen writing in the middle of the third century. See Trigg 1998: 36–37 and 249 (n. 6); and Crouzel 1979 and 1969; with *OCD* 636 (s.v. "Gregory (4) Thaumaturgus") and *ODCC* 713–14 (s.v. "Gregory Thaumaturgus, St.").

427. Lapin 1996: 503.

428. Lapin 1996: 502.

> students of philosophers say about geometry and music and also grammar and rhetoric and astronomy being an assistant to philosophy, we might say the same about philosophy in respect to Christianity.[429]

Origen then goes on to describe an analogy of the Jews despoiling the Egyptians before leaving Egypt. Like them, Christians rob from philosophy what is useful and abandon the rest, "because for some, 'dwelling with the Egyptians', that is to say, with the learning of this world, leads to evil," such as atheism, impiety, and (ironically coming from Origen) heresy.[430] Elsewhere, Origen draws the analogy that philosophy was also akin to the heathen virgins captured by the Hebrew armies in Deuteronomy 21, who could be taken as a concubine if they cut their hair and nails, thus trimming away all that was dead and vain, leaving only what is useful and in agreement with 'acceptable' doctrine.[431] In a different context, he argues that the purpose of natural philosophy for a Christian is to understand the nature of each thing so we do not do anything contrary to nature, so we can know and obey the intentions of the Creator instead, and ultimately learn to "spurn and despise" the natural world and turn instead to the higher things of God. Therefore, to teach us this lesson, Origen claims Solomon established the original three branches of philosophy with his three canonical treatises, *Proverbs* for ethics, the *Song of Songs* for metaphysics and theology, and *Ecclesiastes* for physics.[432]

Here again we have the idea that the Bible is the ultimate science textbook, although Origen at least understood that the Bible had to be interpreted in light of scientific fact, and not the other way around—although that is the attitude that led him to beliefs that were later declared heretical.[433] And

429. Origen, *Letter to Gregory* 1.

430. Origen, *Letter to Gregory* 2–3.

431. Origen, *Homilies on Leviticus* 7.6.6–8.

432. Origen, *Commentary on the Song of Songs* pr.3 (cf. Lawson 1957: 39–46, 317–20).

433. Such as regarding the nature of the resurrection, where in his effort to render it scientifically plausible Origen only guaranteed it would be condemned: cf. Carrier 2005: 123–35, 143–44 (with associated FAQ: www.richardcarrier.info/SpiritualFAQ.html#origen). The same fate befell the similarly science-minded John Philopon in the 6th century: *ODCC* 896 (s.v. "John Philoponus"); *OCD* 1135 (s.v. "Philoponus, John"); *DSB* 7.134–39 (s.v. "John Philoponus"); *NDSB* 4.51–53 (s.v.

Origen did imagine that natural philosophy involved more than a reading of *Ecclesiastes*. Origen thus argues that the Biblical Patriarchs also symbolized the same three branches as Solomon's books: with Isaac representing natural philosophy, because he "dug wells and excavated the depths of things," and that, Origen says, describes natural philosophy, which involves the thorough study of "the nature of each particular thing" and its "causes."[434] So Origen had some idea of studying nature more thoroughly. Yet he still insisted of his students, "devote yourself to reading the divine scriptures first above all."[435] Moreover, he prohibited his students from reading anything written by "godless" scientists or philosophers, which Crouzel rightly identifies as a reference to "les Épicuriens, et à un degré moindre les Péripatéticiens," i.e. "the Epicureans and to a lesser degree the Aristotelians," especially those of a more godless bent, like Strato of Lampsacus (all but forgotten now but then one of the most important scientific authors of the age), though we also saw Justin included the Stoics in this category of condemnation (declaring them in effect "atheists" and thus ruling out the Stoic scholar Posidonius, another widely revered scientific author of the age), and the Skeptics certainly would have been thus condemned, which left only Platonists as

---

"John Philoponus"). Thus the Christian community remained hostile to such openly pro-science attitudes for a long time. Augustine's attempt to revive Origen's notion that scripture cannot refute science (Augustine, *The Literal Interpretation of Genesis* 1.19, cf. Sandnes 2009: 214–30, a remark not as pro-science as it's usually now taken to be: see later note) received only mixed results. Bacon 2001: 5–38 [orig. published 1605] shows the idea still had abundant opposition within the Church even a thousand years later. See P. Harrison 2001 and 1998, Crouch 1975 (esp. pp. 37–90), and Lougee 1972 (esp. pp. 45–60). Their conclusions are supported by Kenny 2004 and 1998; Daston 1998; Eamon 1996; Lloyd 1973: 167–71; and Clagett 1955: 118–82.

434. Origen, *Commentary on the Song of Songs* pr.3 (cf. Lawson 1957: 39–46, 317–20). As for the other two branches, Origen says Abraham represents ethics because of his obedience to God and Jacob represents metaphysics or theology because of his ladder to heaven. Similarly, Origen elsewhere identifies the Biblical Ahuzzath as symbolizing physics because his name means "he who holds," and natural philosophy contains or 'holds' everything in nature (in the context of Origen, *Homilies on Genesis* 14.3, where he first says Abimelech represents logic, and his two subordinates, Ahuzzath and Phicol, represent physics and ethics, respectively).

435. Origen, *Letter to Gregory* 4.

acceptable reading.[436] Which explains why Justin, Origen, and all other early Christian intellectuals looked only to Platonism for ideas to cannibalize for constructing their own Christian philosophy. And indeed extant Christian philosophy in its first three centuries *after* Paul is uniformly most similar to Platonism. Yet Platonism was the least empirical of the pro-science philosophies, still plagued with a hostility to innovative and experimental science, an attitude the Christians appear to have inherited.[437]

The actual aim and content of Origen's curriculum were described by Jerome in the late fourth century:

> In his immortal genius he understood dialectics, as well as geometry, arithmetic, music, grammar, and rhetoric, and taught all the schools of philosophers, in such a way that he had also diligent students in secular literature, and lectured to them daily, and the crowds that flocked to him were marvelous. These he received in the hope that through the instrumentality of this secular literature, he might establish them in the faith of Christ.[438]

Earlier in the fourth century Eusebius confirmed this general picture of Origen's curriculum, noting that Origen gave advanced students a "philosophical education" including "geometry, arithmetic, and other preparatory subjects," and then the major theories of the various philosophical schools, while to less polished students he taught a "grammar school curriculum," all as an aid to the study of scripture.[439] Origen's school thus provided Christians with a godly approach to the full *enkyklios*, followed by a higher education in "Christian-friendly" philosophy, culminating in the life-long study of the scriptures.

---

436. Gregory Thaumaturgus, *Panegyric Oration on Origen* 13 (cf. 11–15). Other schools (like Pythagoreanism) were by that time either nonexistent or too scarcely represented to have mattered educationally, or (like Cynicism) already spurned natural philosophy. On Origen's fondness for Platonism: Edwards 2012.

437. See Carrier 2010; Platonist anti-empiricism of the Roman-era is best exemplified in Plutarch's portrayal of Archimedes (which, though represented as history, is largely fictional, reflecting more the attitudes of Plutarch than Archimedes): Plutarch, *Marcellus* 14.7–12 and 17.5–7. I'll discuss this further in *Scientist*.

438. Jerome, *On Illustrious Men* 54.

439. Eusebius, *History of the Church* 6.18.3.

Origen's students had to have procured an elementary education elsewhere, and we do not know the actual content of Origen's grammar school—it may have simply repeated the pagan model, with occasional sermons against the pagan content of the literature employed. Though many other Christians would have disapproved, Origen was not one to follow the crowd. He had already been kicked out of Alexandria for not kowtowing to the Christian authorities there, and he was ultimately branded a heretic. Origin's program was also just as restricted to the privileged elite as pagan education had been. He also argued that the general public didn't need such schooling, because they had labor and family to devote their time to, and "the full understanding and comprehension" of the facts of the world "will be granted after death" to them anyway.[440] Even apart from that sentiment Origen's small and unique school was available to extremely few Christians. Because it was not emulated elsewhere. Which is a pity, since his program was the most enlightened imaginable for the Christian mindset of the time, including considerable exposure to scientific content, in the form of the *enkyklios* and more advanced natural philosophy. Before his conversion Origen was educated in Platonism, hence again it is no accident his curriculum was primarily Platonic (or that he would be attracted to Christianity as a philosophy—most likely both). So if Plotinus, the leading Platonist educator of Origen's day (whom Origen might well have studied under), could occasionally touch on sciences like optics and mechanics, Origen might have, too.[441] Although his school's rejection of "godless" scientists and philosophers and its emphasis on scripture did not make it entirely science-friendly, it nevertheless anticipated attitudes that would not start to become prevalent (at least in the West) until the late Middle Ages or early Renaissance. The fate of Origen's school after his demise is also telling, since it essentially ended with its last student, Eusebius, only a century later, and we already saw above how Eusebius's negative attitude was not at all what Origen would have hoped for. Indeed, it was apparently the very attitude that put an end to the school itself.

440. Origen, *On the First Principles* 2.11.6–7 (similarly *Against Celsus* 1.9–13 and *On the First Principles* 2.11.4–5).

441. That Plotinus discussed such sciences: Porphyry, *Life of Plotinus* 3, 14, 20; that Origen might have, too: Eusebius, *History of the Church* 6.19. For more on Origen's education and school see Clarke 1971: 125–29; Marrou 1964: 468–69 (= Marrou 1956: 314–15); and Knauber 1968. For a full discussion of Origen's curriculum see Crouzel 1969: 68–70, 141–43, 186–95.

The Christian school at Alexandria may have taught along similar lines, although under tighter dogmatic control from church authorities. We can perhaps get an idea of this from a Christian educator of the early 3rd century, Clement of Alexandria, who *also* commended logic, mathematics, and astronomy, but only in a very qualified and mystical way.[442] Hence he adds that "one must avoid the enormous uselessness of mistaken thinking about irrelevant matters," and so the true Christian:

> Makes use of an education (as much as he finds accessible and not a distraction) as a helpful preparation for the accurate communication of the truth and for a defense against perversions of reason that lead to abandoning the truth. And accordingly, he will not be left deficient in the educational curriculum and philosophy of the Greeks. But they will not be the central object of his concern, but rather a necessary, secondary, and merely circumstantial interest.[443]

By insisting that science and philosophy remain only a circumstantial interest to Christians, Clement seems to treat these subjects as closed and finished, as mere traditions that had only to be passed on, never expanded or improved. Two centuries later, Augustine of Hippo would make a similar point, that Christians ought to familiarize themselves with at least enough science not to look like fools when debating pagans and heretics.[444] But even he never recommends Christians actually *do* science, much less advance it in any way. Neither does Clement.

Indeed, it would seem that Clement cannot even imagine a Christian scientist, or even appears somewhat worried by the idea. To him, science and philosophy must be subordinated to Christian dogma and studied primarily to arm the Christian against attacks upon the faith by scientists and philosophers. Thus, for Clement, science and philosophy take on the role of the enemy's scriptures, a view voiced by other Christians, too.[445]

---

442. Clement of Alexandria, *Stromata* 6.10–11.

443. Clement of Alexandria, *Stromata* 6.10.83. Augustine, *The Literal Interpretation of Genesis* 1.19, 1.20 and 2.9.

444. Augustine, *The Literal Interpretation of Genesis* 1.19, 1.20 and 2.9.

445. For example: Tertullian, *To the Nations* 2.2.42, 2.4.47, *Prescription against Heretics* 7 (cf. 11–14, 21–28), *On the Soul* 1–2; Lactantius, *Divine Institutes* 3; Eusebius, *Preparation for the Gospel* 14.10 (cf. 14.13.9).

Even insofar as philosophy could help one come to Christ, it was to be cast off or subjugated once one arrived there, e.g. Clement cites the scriptural authority of "Solomon," that "when he says, 'Be not much with a strange woman' [Proverbs 5:20], he admonishes us to use indeed, *but not to linger and spend time with*" the philosophy and curriculum of the Greeks.[446] In the end, even Clement's limited approval was not yet representative of any wider Christian sentiment: as Clement himself was aware, most Christians avoided these studies.[447]

From this and all the evidence surveyed above, clearly in its first three centuries Christianity was usually hostile to and always far less supportive of science education than pagans of the same period. Science was not a major educational value for Christians in any sense at all. This must certainly explain much of the stagnation and decline in science and science education after the fourth century. Whatever recovery then occurred in the late Middle Ages and early Renaissance cannot plausibly be attributed to the Christianity that the medievals inherited (having had no such effect for over a thousand years), but can only have resulted from a *change* in Christianity—and it appears that that change consisted in the reintegration of pagan educational values.[448] The evidence of ancient science education therefore offers no support for the thesis that the pre-Christian system of pagan education impeded science, or that the rise of Christianity in any way aided it.

446. Clement of Alexandria, *Stromata* 1.5 (§29.9), where he also draws on Philo (whose views we examined earlier).

447. Clement of Alexandria, *Stromata* 1.9, 6.11.

448. On this point see discussion and sources in Carrier 2010 (esp. pp. 412–19).

10. Conclusion

In the second century A.D. the Epicurean wit Lucian recorded a florid autobiographical account of how he ended up with an education in oratory and philosophy.[449] Though not a natural philosopher, except in the sense that he studied philosophy and possibly the *enkyklios* enough to acquire a lay understanding of the sciences, Lucian tells us how he became a renowned and accomplished writer and orator. His story would not have been typical, but it is unlikely to have been unique. Though highly rhetorical, his account still reflects everything we have surveyed so far and exemplifies probably the furthest range of possibilities normally open to those outside the leisured class.

Lucian says he came from a family of tradesmen—all stoneworkers on his mother's side—and describes himself, somewhat hyperbolically, as the "working-class son of a nobody," destined for a "low-born trade." However, he was still sent to school by his parents all the way into his teens, which means his family had money enough to give him a full primary and secondary education, which already places his family in the minority among the population of the Roman empire (though still solidly middle class). His parents then planned out his subsequent education. Though they ultimately preferred that he follow the family trade by taking an apprenticeship to a stonecutter, they also seriously considered sending him to get more schooling. But they rejected that idea because it would require "a lot of work, a long time, no small amount of money, and remarkable luck," yet they had little money left and needed income fast. This was no doubt

449. For general background see *OCD* 861 (s.v. "Lucian").

a very common situation for working class families like his. So, his family reasoned, if he learned "the art of a craftsman" his apprenticeship would provide his own keep even as he was learning, reducing the burden on his family, while eventually he would graduate and start bringing his own salary home. However, Lucian did not like the work and persuaded his family to send him back to school instead, to gain him a career as an orator.[450] And he clearly did well in that occupation.

Lucian's background was not one of wealth, but not of poverty, either. It was perhaps comparable to the economic situation today of the lower middle class: a family who struggles to make ends meet, but is earning enough to provide an education for one or more of their children, so long as they pinch their pennies and can count on making up their losses on the income of a college graduate. And yet Lucian was able to eek out enough means to afford his way through a full course of higher education. Lucian imagines others could follow in his path, so his circumstances could not have been too unusual, but they still would not have been typical.[451] Even Lucian's family saw nothing bizarre in having educated their son only to prepare him for the ancient equivalent of a blue-collar career, which implies literate craftsmen were not unusual, either, even if again not typical. But besides completing and mastering primary, secondary, and rhetorical school, Lucian went on to study philosophy, declaring passionate allegiance to the Epicurean sect, yet he shows signs of enough familiarity with other sects to suggest he had studied under several, which *was* typical for a philosophy student. Though he never says how he afforded this education or whether he persuaded his family to underwrite his expenses, he clearly managed it, one way or the other—possibly by attaching himself as a client to a wealthy patron. And though very few, even among those who had adequate means, would have had such luck, initiative, talent, and drive, Lucian was still not alone in these respects, either.

The reality for most Romans, however, is captured more generally in the late third century, by the Christian educator Lactantius, advisor to the emperor Constantine, who confirms that few advanced to higher education

---

450. Lucian, *On the Dream* or *Lucian's Career* 1, 7, 10.

451. Lucian, *On the Dream* or *Lucian's Career* 11, 18. The poet Horace may have had a similar story, his father financing his extensive education with an aim at social advancement (see *OCD* 704–07, in s.v. "Horace (Quintus Horatius Flaccus)").

as Lucian did. Speaking of the many idealistic philosophers in antiquity who argued that everyone should get an education—men and women, slave or free—Lactantius says:

> They attempted, indeed, to do that which truth required, but they were unable to proceed beyond words. First, because instruction in many arts is necessary for an application to philosophy. Common learning must be acquired on account of practice in reading, because in so great a variety of subjects it is impossible that all things should be learned by hearing, or retained in the memory. No little attention also must be given to the grammarians, in order that you may know the right method of speaking. That must occupy many years. Nor must there be ignorance of rhetoric, that you may be able to utter and express the things which you have learned. Geometry also, and music, and astronomy, are necessary, because these arts have some connection with philosophy; and the whole of these subjects cannot be learned by women, who must learn within the years of their maturity the duties which are thereafter soon to be of service to them for domestic life; nor by servants, who must live in service during those years especially in which they are able to learn; nor by the poor, or laborers, or rustics, who have to gain their daily support by labor.
>
> On this account Cicero says that philosophy is absent from the multitude. True, Epicurus will receive the uneducated. But how will they understand those things which are said respecting the first principles of things, the perplexities and intricacies of which are scarcely attained to by men of cultivated minds? Therefore, in subjects which are involved in obscurity, and confused by a variety of intellects, and set off by the studied language of eloquent men, what place is there for the unskilled and ignorant?[452]

Lactantius thus sums up the reality of ancient education, including the difficulty of access to scientific content. For Lactantius, as we saw in the previous chapter, this was actually grounds for rejecting science and philosophy, for surely God would not make wisdom difficult or inaccessible to the common man. Therefore only simple, oral learning can be at all worthwhile—as long as it was suitably divine, as the Gospel was, hence "that alone is wisdom."[453] As we saw in the last chapter, this was a common Christian attitude of the time.

---

452. Lactantius, *Divine Institutes* 3.25.

453. Lactantius, *Divine Institutes* 3.26–27.

Yet Galen arrives at exactly the opposite conclusion from the same general observations:

> People would do well to find others to give an assessment of their own opinions. But the people they go to should not be of the same stamp as themselves, that is, untrained in the methods of logical proof, as well as in the other subjects (geometry, mathematics, arithmetic, engineering, astronomy) by which the soul is sharpened. Some people have not even enjoyed the schooling of an orator, or for that matter of a grammarian, which is the most widely available sort of education of all. They are so completely lacking in any sort of verbal training that they cannot follow the arguments they hear from my lips. When making a speech I sometimes notice that this is the case, and ask them to repeat what I have just said, for it is plain that they are like asses listening to music—completely unable to follow the sense of my words. Nonetheless, their arrogance or cheek is such that even when subjected to the open scorn of persons who are literate, for their inability to give an account of the arguments they have just heard, they experience no shame, but actually believe that the truth is known to them alone, and that those who have bothered to educate themselves have merely wasted their time.
>
> But it is no part of the purpose of my argument here to attempt the salvation of such people as those—a salvation that would be impossible for far better men than them, even assuming that they desired to be saved. For they are not at the age which lends itself to education. My work here, I hope, will be of assistance to the man of natural intelligence who also had that early training which gives him the ability, preferably, to repeat immediately whatever argument he hears, and at least to write it down. He must, additionally, be completely dedicated to the pursuit of the truth—this last condition depends entirely on him. The first requirement, though, is that of the right natural endowments for the pursuit of truth, and the second that of a decent early education. One who is not so endowed by nature, and who has been brought up to bad, licentious habits, will never have that desire for the truth, neither from his own personal impulses nor from the encouragement of others. I myself have never claimed to be able to assist such a person. As I have said, I can only help the man who is a friend of truth.[454]

454. Galen, *On the Affections and Errors of the Soul* 2.2 (= Kühn 5.64–66). Galen composed an abbreviated version of this same point in *On the Natural Faculties* 3.10 (= Kühn 2.178–80). For similar arguments elsewhere see Galen, *On the Doctrines of Hippocrates and Plato* 2.3.12–17 and *On the Uses of the Parts* 10.12, 10.14, 12.6 (= May 1968: 490, 502, 558–60).

This confirms several things: the presence of illiterates and the poorly educated even in the audiences of lecturing scientists; the wide availability of the trivium for those who could manage it; and the value of the quadrivium as well as the rarity of those who completed it. But overall, the point of Galen's treatise quoted above, is that bad character can only be reformed by living a life of honest self-reflection and study, which requires a good grasp of logic aided by completion of the *enkyklios*, which is exactly opposite the argument of Lactantius, who claims that approaches like Galen's are useless and never work, but that simple expressions of God's wisdom, a brief reading or two from scripture, a little bit about heaven and hell, quickly reforms even the worst characters. Give him any moral reprobate, Lactantius says, and "with a very few words of God I will render him as gentle as a sheep."[455] The Gospel is all powerful, while anything that actually requires an education is useless and ineffective. These two men are poles apart in fundamental values. And yet they are representative of their respective sides of the ancient culture-war between Christians and pagans in the early Roman empire.[456]

Nevertheless, Galen and Lactantius agree on one thing: to gain any significant exposure to scientific knowledge and scientific reasoning, one first had to climb the steep and expensive hill of basic education, at both the primary and secondary levels, mastering first the various skills of reading and writing and basic arithmetic, and then still have the time and money—and proven ability—to study the quadrivium (of advanced arithmetic, geometry, music theory, and astronomy), even if you planned to skip an advanced rhetorical education and go directly to the philosophy schools. This made the study of science even harder and more off-putting than it might already have been. It required an exceptional degree of money, discipline, commitment, and hard work even to get to the point where studying any science under any significant expert would be possible (although that remains almost as true even to this day).[457] And though Tacitus, Quintilian, and Cicero all said a rhetor's education would benefit greatly from a well-rounded education in science and natural philosophy, this was not the focus

455. Lactantius, *Divine Institutes* 3.26.

456. This culture war has been examined in regard to pagan and Christian attitudes toward miracles in Grant 1952 and Remus 1983; and in other respects by Fox 1987 and MacMullen 1984 and 1997.

457. So Cribiore 2001: 1–4, 220–23.

of a rhetorical education, and many schools and students no doubt failed to achieve their ideal. Only the wealthiest or most dedicated students and those pursuing practical or academic careers in philosophy or science were likely to learn very much. And even then, the quality of information anyone learned would vary, and many students would get a good dose of nonsense along with reasonable facts and theories. Though it bears repeating, it's doubtful things were any different in the Middle Ages. Indeed, the situation was likely worse.

We have also seen there was always less social prestige, or at any rate less social interest, in scientific research than in literary achievement and displays of erudition—accomplishments derived more from rhetorical schooling than anything like a scientific or philosophical education. But there was, nevertheless, enough prestige and interest in science to fuel its advance. Though science and natural philosophy had not become a primary social value at the level of educational standards in antiquity, it was not slighted either. Indeed, it is sometimes claimed that the Romans were more interested in writing reference books and encyclopedias in the sciences than in doing original research.[458] But there is no evidence of a lack of research. In fact, though there is evidence of a growing fad for reference books under the Romans, this would demonstrate a *rising* interest in the sciences, not a decline (as the multiplication of science documentaries on modern television similarly reflects), although whether there even was a rise is questionable, due, again, to Medieval, not ancient, choices as to which books to preserve.[459] But the same values were represented even in Classical and Hellenistic encyclopedic work. Only when society had abandoned almost all serious interest in scientific research and the epistemic values necessary to it, as it did after the third century A.D., did the writing of reference books *replace* scientific research.[460] However, until that happened, ancient

---

458. For example: Diederich 1999: 66–67.

459. See Witty 1974 (although incomplete and outdated). On encyclopedism as in fact a pre-Roman fad beginning in the very heyday of Alexandrian science, see two chapters on the subject in König & Woolf 2013: 23–83. On the phenomenon in the Roman period: Doody 2009.

460. The continued advance of scientific research through the early Roman Empire is briefed in Carrier 2010; it's abandonment thereafter becomes clear by comparison. I will demonstrate this in *The Scientist in the Early Roman Empire*.

education worked well enough to fuel a slow pace of scientific progress, numerous scientific professionals received remarkably good educations, and even many laypersons came to be reasonably well educated in scientific subjects.

There remained an enormous difference between Christian and pagan ideals regarding the value of science education and the extent of its pursuit among the economic and intellectual elite, but that would not impact society until after Constantine. Though even under pagan tenure the relatively low status of science in ancient education ensured that scientific knowledge and values would always be fragile (and thus they were shattered by the decline of the empire and the subsequent triumph of Christianity), this did not signify anything like a negative or hostile attitude from the dominant elite of the time, or even indifference (as science had pervasive respect), nor did it prevent the successful pursuit and advancement of the sciences. Any comparison with the Middle Ages must take all these facts into account.

Bibliography

Abbreviations commonly used in notes:

DSB = Gillispie 1980
EANS = Keyser & Irby-Massie 2008
Kühn = Kühn 1821–1833[461]
LSG = Liddell & Scott 1996
LSL = Lewis & Short 1879
NDSB = Koertge 2008
OCD = Hornblower & Spawforth 2012
ODCC = Cross & Livingstone 1997

Adams, James Noel. 2003. *Bilingualism and the Latin Language*. New York: Cambridge University Press.

Adams, James Noel, Mark Janse and Simon Swain, eds. 2002. *Bilingualism in Ancient Society: Language Contact and the Written Text*. Oxford: Oxford University Press.

Agusta-Boularot, Sandrine. 2004. "Les femmes, l'éducation et l'enseignement dans le monde romain depuis le livre d'H.-I. Marrou." [in Pailler & Payen 2004: 319–30]

461. There is no consistent numbering system for passages in Galen other than (in most cases) Kühn 1821–1833, which I give whenever possible. If I provide any other numeration it will follow the scheme used in the most recent English translation prior to 2008.

Amundsen, Darrel. 1978. "The Forensic Role of Physicians in Ptolemaic and Roman Egypt." *Bulletin of the History of Medicine* 52.3 (Fall): 336–53.

———. 1979. "The Forensic Role of Physicians in Roman Law." *Bulletin of the History of Medicine* 53.1 (Spring): 39–56.

Anderson, Graham. 1993. *The Second Sophistic: A Cultural Phenomenon in the Roman Empire*. London: Routledge.

Asmis, Elisabeth. 2004. "L'éducation épicurienne." [in Pailler & Payen 2004: 211–18]

Asper, Markus. 2007. *Griechische Wissenschaftstexte: Formen, Funktionen, Differenzierungsgeschichten*. Stuttgart: F. Steiner Verlag.

Atkins, Margaret, and Robin Osborne, eds. 2006. *Poverty in the Roman World*. Cambridge: Cambridge University Press.

Bacon, Francis. 2001 [orig. 1605]. *The Advancement of Learning*. Stephen Jay Gould, ed. New York: Modern Library.

Bagnall, Roger, ed. 2009. *The Oxford Handbook of Papyrology*. New York: Oxford University Press.

Ballér, Piroska. 1992. "Medical Thinking of the Educated Class in the Roman Empire: Letters and Writings of Plutarch, Fronto and Aelius Aristides." *From Epidaurus to Salerno: Symposium Held at the European University Centre for Cultural Heritage, Ravello, April, 1990*. Antje Krug, ed. Rixensart: PACT Belgium: 19–24.

Barker, Andrew. 1989. *Greek Musical Writings II: Harmonic and Acoustic Theory*. Cambridge: Cambridge University Press.

———. 1994. "Greek Musicologists in the Roman Empire." [in T. Barnes 1994: 53–74]

Barker, Peter, and Bernard Goldstein. 1984. "Is Seventeenth Century Physics Indebted to the Stoics?" *Centaurus* 27.2: 148–64.

Barnes, Jonathan. 1988. "Scepticism and the Arts." *Apeiron: A Journal for Ancient Philosophy and Science* 21.2 (Summer): 53–77.

———, ed. 1995. *The Cambridge Companion to Aristotle*. New York: Cambridge University Press.

———. 1997. *Logic and the Imperial Stoa*. New York: Brill.

———. 2002. "Ancient Philosophers." [in Clark and Rajak 2002: 293–306]

Barnes, Timothy, ed. 1994. *The Sciences in Greco-Roman Society*. Edmonton, Alberta: Academic. [= *Apeiron: A Journal for Ancient Philosophy and Science* 27.4 (December)]

Barton, Tamsyn. 1994a. *Power and Knowledge: Astrology, Physiognomics, and Medicine under the Roman Empire*. Ann Arbor: University of Michigan Press.

———. 1994b. *Ancient Astrology*. New York: Routledge.

Bauman, Richard. 1992. "Women in Law." *Women and Politics in Ancient Rome*. London: Routledge: 45–52.

Beagon, Mary. 1992. *Roman Nature: The Thought of Pliny the Elder*. Oxford: Clarendon Press.

Beaujouan, Guy. 1963. "Motives and Opportunities for Science in the Medieval Universities." [in Crombie 1963: 219–36]

Ben-David, Joseph. 1984. *The Scientists Role in Society: A Comparative Study* [with a new introduction]. Chicago: University of Chicago Press.

"Bibliotheken." 1897. *Paulys Real-Encyclopädie der classischen Altertumswissenschaft* 5: 405–24.

Bicknell, Peter. 1983. "The Witch Algaonice and Dark Lunar Eclipses in the Second and First Centuries BC." *Journal of the British Astronomical Association* 93: 160–63.

Bloomer, W. Martin. 2011. *The School of Rome: Latin Studies and the Origins of Liberal Education*. Berkeley: University of California Press.

Blum, Rudolf. 1991. *Kallimachos: The Alexandrian Library and the Origins of Bibliography*. Madison, Wisconsin: University of Wisconsin Press.

Boatwright, Mary. 1987. *Hadrian and the City of Rome*. Princeton: Princeton University Press.

———. 2000. *Hadrian and the Cities of the Roman Empire*. Princeton: Princeton University Press.

Bobzien, Susanne. 2011. "The Combinatorics of Stoic Conjunction: Hipparchus Refuted, Chrysippus Vindicated." *Oxford Studies in Ancient Philosophy* 40: 157–88.

Bonner, Stanley. 1977. *Education in Ancient Rome: From the Elder Cato to the Younger Pliny*. Berkeley: University of California Press.

Bowen, Alan, and Robert Todd. 2004. *Cleomedes' Lectures on Astronomy: A Translation of* The Heavens *with an Introduction and Commentary*. Berkeley: University of California Press.

Bowen, James. 1972. *A History of Western Education. Vol. 1, The Ancient World: Orient and Mediterranean, 2000 B.C. - A.D.1054*. London: Methuen.

———. 1975. *A History of Western Education. Vol. 2, Civilization of Europe, Sixth to Sixteenth Century*. London: Methuen, 1975.

Bowersock, G.W. 1969. *Greek Sophists in the Roman Empire*. Oxford: Clarendon Press.

———. 1974. *Approaches to the Second Sophistic: Papers Presented at the 105th Annual Meeting of the American Philological Association*. University Park, Pennsylvania: American Philological Association.

———. 2002. "Philosophy in the Second Sophistic." [in Clark and Rajak 2002: 157–70]

Boyd, Clarence Eugene. 1915. *Public Libraries and Literary Culture in Ancient Rome*. Chicago: University of Chicago Press.

Boylan, Michael. 1983. *Method and Practice in Aristotle's Biology*. Washington, D.C.: University Press of America.

Brenk, Frederick E. 2007. "School and Literature: The 'Gymnasia' at Athens in the First Century A.D." *Escuela y literatura en Grecia antigua: Actas del simposio internacional: Universidad de Salamanca, 17–19 noviembre de 2004*. José Antonio Fernández Delgado, Francisca Pordomingo, Antonio Stramaglia, eds. Cassino: Ed. dell'Università degli Studi di Cassino: 333–347.

Brodie, Thomas. 2004. *The Birthing of the New Testament: The Intertextual Development of The New Testament Writings*. Sheffield: Sheffield Phoenix Press.

Broek, Roelof van den. 1995. "The Christian 'School' of Alexandria in the Second and Third Centuries." *Centres of Learning: Learning and Location in Pre-Modern Europe and the Near East*. Jan Willem Drijvers and Alasdair MacDonald, eds. Leiden: E.J. Brill: 39–47.

Brunschwig, Jacques. 2002. "Cicero, Zeno of Citium, and the Vocabulary of Philosophy." *Le style de la pensée: Recueil de textes en hommage*. Monique Canto-Sperber and Pierre Pellegrin, eds. Paris: Belles Lettres: 412–28.

———. 2007. "Science et philosophie chez les stoïciens." *Conceptions de la science: hier, aujourd'hui, demain: hommage à Marjorie Grene*. Jean Gayon et al., eds. Brussels: Ousia: 73–91.

Brunt, P.A. 1994. "The Bubble of the Second Sophistic." *Bulletin of the Institute of Classical Studies* 39: 25–52.

Byl, Simon. 1997. "Controverses antiques autour de la dissection et de la vivisection." *Revue Belge de Philologie et d'Histoire* 75.1: 113–20.

Campbell, Brian. 2000. *The Writings of the Roman Land Surveyors: Introduction, Text, Translation and Commentary*. Society for the Promotion of Roman Studies.

Canfora, Luciano. 1987. *The Vanished Library: A Wonder of the Ancient World*. Berkeley: University of California Press. [tr. by Martin Ryle]

Cao, Irene. 2010. Alimenta: *il racconto delle fonti*. Padova: Il Poligrafo.

Carrier, Richard. 2003. "Only a Fraction Was Literate." *Biblical Archaeology Review* 29.6 (November/December): 13.

———. 2005. "The Spiritual Body of Christ and the Legend of the Empty Tomb." *The Empty Tomb: Jesus Beyond the Grave*. Robert M. Price and Jeffery Jay Lowder, eds. Amherst, New York: Prometheus: 105–232.

———. 2010. "Christianity Was Not Responsible for Modern Science." *The Christian Delusion: Why Faith Fails*. John Loftus, ed. Amherst, New York: Prometheus: 396–419.

———. 2011. "Christianity's Success Was Not Incredible." *The End of Christianity*. John Loftus, ed. Amherst, New York: Prometheus Books.

———. 2014. *On the Historicity of Jesus: Why We Might Have Reason for Doubt*. Sheffield, UK: Sheffield-Phoenix Press.

Carriker, Andrew. 2003. *The Library of Eusebius of Caesarea*. Leiden: Brill.

Casey, Eric. 2014. "Educating the Youth: The Athenian *Ephebeia* in the Early Hellenistic Era." *The Oxford Handbook of Childhood and Education in the Classical World*. Judith Evans Grubbs and Tim G. Parkin, eds. New York: Oxford University Press: 418–43.

Casson, Lionel. 2001. *Libraries in the Ancient World*. New Haven: Yale University Press.

Catana, Leo. 2013. "The Origin of the Division between Middle Platonism and Neoplatonism." *Apeiron* 46.2: 166–200.

Chankowski, Andrzej. 2004. "L'éphébie à l'époque hellénistique: institution d'éducation civique." [in Pailler & Payen 2004: 271–80]

Chapman, Paul. 2001. "The Alexandrian Library: Crucible of a Renaissance." *Neurosurgery* 49.1 (July): 1–13.

Cheyne, James Allen. 2010. "Atheism Rising: The Connection between Intelligence, Science, and the Decline of Belief." *Skeptic* 15.2: 33–41.

Chiaradonna, Riccardo, and Franco Trabattoni. 2009. *Physics and Philosophy of Nature in Greek Neoplatonism: Proceedings of the European Science Foundation Exploratory Workshop (Il Ciocco, Castelvecchio Pascoli, June 22–24, 2006)*. Leiden: Brill.

Christes, Johannes. 1988. "Gesellschaft, Staat und Schule in der griechisch-römischen Antike." *Sozialmassnahmen und Fürsorge*. Hans Kloft, ed. Graz: Horn: 55–74.

Christianidis, Jean, and Jeffrey Oaks. 2013. "Practicing Algebra in Late Antiquity: The Problem-Solving of Diophantus of Alexandria." *Historia Mathematica: International Journal of the History of Mathematics* 40.2: 127–63.

Clagett, Marshall. 1955. *Greek Science in Antiquity*. Salem, New Hampshire: Ayer.

Clark, Gillian and Tessa Rajak, eds. 2002. *Philosophy and Power in the Graeco-Roman World: Essays in Honour of Miriam Griffin* (New York: Oxford University Press).

Clarke, M.L. 1971. *Higher Education in the Ancient World*. Albuquerque, New Mexico: University of New Mexico Press.

Clarysse, Willy, and Dorothy Thompson. 2006. *Counting the People in Hellenistic Egypt*. 2 vols. Cambridge: Cambridge University Press.

Cohen, Ada, and Jeremy Rutter, eds. 2007. *Constructions of Childhood in Ancient Greece and Italy*. Princeton, New Jersey: American School of Classical Studies at Athens.

Cohen, Morris, and I.E. Drabkin. 1948. *A Source Book in Greek Science*. London: Oxford University Press.

Cohn-Haft, L. 1956. *The Public Physicians of Ancient Greece*. Northampton, Massachusetts: Smith College Publications in History.

Collins, Randall. 1998. *The Sociology of Philosophies: A Global Theory of Intellectual Change*. Cambridge, Massachusetts: Harvard University Press.

Connolly, R. Hugh. 1929. *Didascalia Apostolorum*. Oxford: Clarendon Press.

Constantelos, Demetrios. 1998. *Christian Hellenism: Essays and Studies in Continuity and Change*. New Rochelle, New York: Aristide D. Caratzas.

Copan, Paul. 1998. "Augustine and the Scandal of the North African Catholic Mind." *Journal of the Evangelical Theological Society* 41.2 (June): 287–95.

Cribiore, Raffaella. 1996. *Writing, Teachers, and Students in Graeco-Roman Egypt*. Atlanta: Scholars Press.

———. 2001. *Gymnastics of the Mind: Greek Education in Hellenistic and Roman Egypt*. Princeton, N.J.: Princeton University Press.

———. 2007. "Lucian, Libanius, and the Short Road to Rhetoric." *Greek, Roman, and Byzantine Studies* 47.1: 71–86.

———. 2009. "The Education of Orphans: A Reassessment of the Evidence of Libanius." *Growing up Fatherless in Antiquity*. Sabine Hübner and David Ratzan, eds. New York: Cambridge University: 257–72.

Crivelli, Paolo. 2007. "Epictetus and Logic." *The Philosophy of Epictetus*, ed. Thedore Scaltsas and Andrew Mason. New York: Oxford University Press: 20–31.

Crombie, A.C. 1963. *Scientific Change: Historical Studies in the Intellectual, Social, and Technical Conditions for Scientific Discovery and Technical Invention, from Antiquity to the Present (Symposium on the History of Science, Oxford, 9–15 July 1963)*. New York: Basic Books.

Cross, F.L., and E.A. Livingstone, eds. 1997. *The Oxford Dictionary of the Christian Church*, 3rd ed. Oxford: Oxford University Press.

Crouch, Laura. 1975. *The Scientist in English Literature: Domingo Gonsales (1638) to Victor Frankenstein (1817)*. Dissertation, Ph.D. (University of Oklahoma).

Crouzel, Henri. 1969. *Remerciement à Origène Suivi de la Lettre d'Origène à Grégoire*. Paris: Cerf.

————. 1979. "Faut-il voir trois personnages en Grégoire le Thaumaturge?" *Gregorianum* 60: 287–320.

Cuomo, Serafina. 2000. *Pappus of Alexandria and the Mathematics of Late Antiquity*. Cambridge: Cambridge University Press.

————. 2001. *Ancient Mathematics*. London: Routledge.

————. 2012. "Exploring Ancient Greek and Roman Numeracy." *BSHM Bulletin: Journal of the British Society for the History of Mathematics* 27: 1–12.

Daston, Lorraine. 1998. *Wonders and the Order of Nature, 1150–1750*. New York: Zone Books.

Davies, R. W. 1970. "The Roman Military Medical Service." *Saalburg-Jahrbuch* 27: 84–104.

Deakin, Michael. 2007. *Hypatia of Alexandria: Mathematician and Martyr*. Amherst: Prometheus Books.

Debru, Armelle. 1995. "Les démonstrations médicales à Rome au temps de Galien." [in van der Eijk, Horstmanshoff and Schrijvers 1995: 1.69–82]

Delorme, Jean. 1960. *Gymnasion: Étude sur les monuments consacrés a l'education en Gréce (des origines à l'Empire Romain)*. Paris: Éditions E. De Boccard.

Demont, Paul. 2004. "H.-I. Marrou et «les deux colonnes du temple»: Isocrate et Platon." [in Pailler & Payen 2004: 109–19]

Derbyshire, John. 2006. *Unknown Quantity: A Real and Imaginary History of Algebra*. Washington, D.C.: Joseph Henry Press.

De Ridder-Symoens, Hilde, ed. 1992. *A History of the University in Europe*, Vol. 1. Cambridge: Cambridge University.

Desmond, William. 2008. *Cynics*. Stocksfield: Acumen.

DeVoto, James. 1996. *Philon and Heron: Artillery and Siegecraft in Antiquity*. Chicago: Ares.

Diederich, Silke. 1999. "Zur Rezeption der Naturwissenschaften in der römischen Schule der Kaiserzeit." *Antike Naturwissenschaft und ihre Rezeption* 9: 45–68.

Dilke, O.A.W. 1971. *The Roman Land Surveyors: An Introduction to the* Agrimensores. Newton Abbot: David & Charles.

Dillon, John. 1993. *Alcinous: The Handbook of Platonism*. Oxford: Clarendon.

Dillon, John, and A.A. Long, eds. 1988. *The Question of 'Eclecticism': Studies in Later Greek Philosophy*. Berkeley: University of California Press.

Di Muzio, Gianluca. 2007. "Epicurus' Emergent Atomism." *Philo* 10.1 (Spring-Summer): pp. 5–16.

Dobson, J.F. 1932. *Ancient Education and its Meaning to Us*. New York: Cooper Square.

Donderer, Michael. 1996. *Die Architekten der späten römischen Republik und der Kaiserzeit*. Erlangen: Universitätsbund Erlangen-Nürnberg.

Doody, Aude. 2009. "Pliny's *Natural History*: *Enkuklios Paideia* and the Ancient Encyclopedia." *Journal of the History of Ideas* 70.1: 1–21.

Doody et al. 2012. "Structures and Strategies in Ancient Greek and Roman Technical Writing." *Studies in History and Philosophy of Science* 43 (Special Issue).

D'Ooge, Martin Luther, Frank Egleston Robbins and Louis Charles Karpinski. 1926. *Nicomachus of Gerasa: Introduction to Arithmetic*. New York: Macmillan.

Duckworth, W.L.H., M..C. Lyons and B. Towers. 1962. *Galen On Anatomical Procedures: The Later Books*. 1962. Cambridge: Cambridge University Press.

Dufallo, Basil. 2005. "Words Born and Made: Horace's Defense of Neologisms and the Cultural Poetics of Latin." *Arethusa* 38.1: 89–101.

Duffy, John. 1984. "Byzantine Medicine in the 6th and 7th Centuries: Aspects of Teaching and Practice." *Dumbarton Oaks Papers* 38: 21–27.

Duncan-Jones, Richard P. 1982. *The Economy of the Roman Empire: Quantitative Studies*. 2nd ed. Cambridge: Cambridge University Press.

Durling, Richard. 1995. "Medicine in Plutarch's Moralia." *Traditio: Studies in Ancient and Medieval History, Thought, and Religion* 50: 311–14.

Dutch, Robert. 2005. *The Educated Elite in 1 Corinthians: Education and Community Conflict in Graeco-Roman Context*. London: T & T Clark.

Dzielska, Maria. 1995. *Hypatia of Alexandria*. Cambridge, Massachusetts: Harvard University Press.

Eamon, William. 1996. *Science and the Secrets of Nature*. Princeton: Princeton University Press.

Edelstein, Ludwig. 1952. "Recent Trends in the Interpretation of Ancient Science." *Journal of the History of Ideas* 13.4 (October): 573–604.

———. 1963. "Motives and Incentives for Science in Antiquity." [in Crombie 1963: 15–41]

———. 1967. *The Idea of Progress in Classical Antiquity*. Baltimore: Johns Hopkins Press.

Edwards, Mark. 2012. "Further Reflections on the Platonism of Origen." *Adamantius* 18: 317–24.

Efron, Noah. 2009. "Myth 9: That Christianity Gave Birth to Modern Science." *Galileo Goes to Jail: And Other Myths about Science and Religion*, ed. Ronald Numbers. Cambridge, Mass.: Harvard University Press: 79–89.

Eijk, Philip J. van der, H. Horstmanshof, and P. Schrijvers. 1995. *Ancient Medicine in its Socio-Cultural Context: Papers Read at the Congress Held at Leiden University, 13–15 April 1992*. Atlanta, GA: Rodopi. [= *Clio Medica* 27 and 28]

El-Abbadi, Mostafa. 1992. *Life and Fate of the Ancient Library of Alexandria*, 2nd. ed., rev. Paris: United Nations Educational, Scientific and Cultural Organization (UNESCO).

El-Abbadi, Mostafa and Omnia Mounir Fathallah, eds. 2008. *What Happened to the Ancient Library of Alexandria?* Boston: Brill.

Ellspermann, Gerard. 1949. *The Attitude of the Early Christian Latin Writers toward Pagan Literature and Learning*. Washington, D.C.: Catholic University of America Press.

Engberg-Pedersen, Troels. 2000. *Paul and the Stoics*. Louisville, Ky.: Westminster John Knox Press.

———. 2010. *Cosmology and Self in the Apostle Paul: The Material Spirit*. Oxford : Oxford University Press.

Evans, Edith. 1994. "Military Architects and Building Design in Roman Britain." *Britannia* 25: 143–64.

Evans, James. 1999. "The Material Culture of Greek Astronomy." *Journal for the History of Astronomy* 30.3 (August): 237–307.

Evans, James, and J. L. Berggren. 2006. *Geminos's Introduction to the Phenomena: A Translation and Study of a Hellenistic Survey of Astronomy*. Princeton: Princeton University Press.

Eyre, J.J. 1963. "Roman Education in the Late Republic and Early Empire." *Greece & Rome* 10.1 (March): 47–59.

Falcon, Andrea. 2013. "Aristotelianism in the First Century B.C.: Xenarchus of Seleucia." *Aristotle, Plato and Pythagoreanism in the First Century B.C.: New Directions for Philosophy*. Malcolm Schofield, ed. Cambridge: Cambridge University Press: 78–94.

Fehrle, R. 1986. *Das Bibliothekswesen in alten Rom: Voraussetzungen, Bedingungen, Anfänge*. Wiesbaden, Universitätsbibliothek.

Ferguson, John, and Jackson P. Hershbell. 1990. "Epicureanism under the Roman Empire." *Aufstieg und Niedergang der römischen Welt* 2.36.4: 2257–2327.

Ferngren, Gary. 1982. "A Roman Declamation on Vivisection." *Transactions & Studies of the College of Physicians of Philadelphia* 4.4 (December): 272–90.

Ferruolo, Stephen. 1998. *The Origins of the University: The Schools of Paris and their Critics, 1100–1215*. Stanford: Stanford University.

Flemming, Rebecca. 2007. "Women, Writing and Medicine in the Classical World." *The Classical Quarterly* 57.1 (May): 257–279.

Fögen, Thorsten. 2000. *Patrii Sermonis Egestas: Einstellungen lateinischer Autoren zu ihrer Muttersprache: ein Beitrag zum Sprachbewusstsein in der römischen Antike*. Münich: K.G. Saur.

Forbes, Clarence. 1945. "Expanded Uses of the Greek Gymnasium." *Classical Philology* 40.1 (January): 32–42.

Fox, Robin Lane. 1987. *Pagans and Christians*. New York: Alfred A. Knopf.

Freeman, Charles. 2002. *The Closing of the Western Mind: The Rise and Fall of Reason*. London: Heinemann.

French, Roger, and Frank Greenaway, eds. 1986. *Science in the Early Roman Empire: Pliny the Elder, His Sources and Influence*. London: Croom Helm.

Furley, David, and J.S. Wilkie. 1984. *Galen on Respiration and the Arteries*. Princeton University Press.

Gain, D.B. 1976. *The Aratus Ascribed to Germanicus Caesar*. London: Athlone Press.

Gamble, Harry. 1995. *Books and Readers in the Early Church: A History of Early Christian Texts*. New Haven: Yale University Press.

Gauthier, Philippe. 1995. "Notes sur le rôle di gymnase dans les cités hellénistiques." *Stadtbild und Bürgerbild im Hellenismus: Kolloquium, München, 24. bis 26. Juni 1993*. Michael Worrle and Paul Zanker, eds. Munich: Beck: 1–11.

Gee, Emma. 2013. *Aratus and the Astronomical Tradition*. Oxford: Oxford University Press.

Gemeinhardt, Peter. 2012. "In Search of Christian *paideia*: Education and Conversion in Early Christian Biography." *Zeitschrift für antikes Christentum* 16.1: 88–98.

Georges, Tobias. 2012. "Justin's school in Rome: Reflections on Early Christian 'schools.'" ». *Zeitschrift für antikes Christentum* 16.1: 75–87.

Gerhardsson, Birger. 1961. *Memory and Manuscript: Oral Tradition and Written Transmission in Rabbinic Judaism and Early Christianity*. Lund: C.W.K. Gleerup.

Gerson, Lloyd. 2013. *From Plato to Platonism*. Ithaca: Cornell University Press.

Gibson, Craig. 2013. "Doctors in Ancient Greek and Roman Rhetorical Education." *Journal of the History of Medicine & Allied Sciences* 68.4 (October): 529–50.

Gillispie, Charles Coulston, ed. 1980. *Dictionary of Scientific Biography*. Vols. I-XVI. New York: Charles Scribner's Sons.

Goguey, D. 1978. "La formation de l'architecte: culture et technique." *Recherches sur les 'artes' à Rome*, ed. Jean-Marie André. Paris: Les Belles Lettres: 100–115.

Gottschalk, H.B. 1987. "Aristotelian Philosophy in the Roman World from the Time of Cicero to the End of the Second Century A.D." *Aufstieg und Niedergang der römischen Welt* 2.36.2: 1079–1174.

Götze, Bernt. "Antike Bibliotheken" *Jahrbuch des Deutschen Archäologischen Instituts* 52 (1937): 223–247.

Gourevitch, Danielle. 1970. "Some Features of the Ancient Doctor's Personality as Depicted in Epitaphs." *Nordisk Medicinhistorisk Årsbok* 1970: 38–49.

Grant, Robert M. 1952. *Miracle and Natural Law in Graeco-Roman and Early Christian Thought*. Amsterdam: North-Holland.

Green, Peter. 1990. *Alexander to Actium: The Historical Evolution of the Hellenistic Age*. Berkeley: University of California Press.

Greene, Kevin. 1992. "How Was Technology Transferred in the Western Provinces?" *Current Research on the Romanization of the Western Provinces*. Mark Wood and Francisco Queiroga, eds. Oxford: Tempvs Reparatvm: 101–05.

————. 1994. "Technology and Innovation in Context: The Roman Background to Mediaeval and Later Developments." *Journal of Roman Archaeology* 7: 22–33.

Gruen, Erich. 1998. *Heritage and Hellenism: The Reinvention of Jewish Tradition*. Berkeley: University of California Press.

————. 2002. *Diaspora: Jews amidst Greeks and Romans*. Cambridge, Mass.: Harvard University Press.

Gruner, Rolf. 1975. "Science, Nature, and Christianity." *The Journal of Theological Studies* 26: 55–81.

Gunderson, Erik, ed. 2009. *The Cambridge Companion to Ancient Rhetoric*. New York: Cambridge University Press.

Gwynn, Aubrey. 1926. *Roman Education from Cicero to Quintilian*. New York: Russell & Russell. [citing a 1964 reprint]

Haarhoff, Theodore. 1920. *Schools of Gaul: A Study of Pagan and Christian Education in the Last Century of the Western Empire*. London: Oxford University Press.

Haines-Eitzen, Kim. 2000. "Learning to Write in Graeco-Roman Antiquity." *Guardians of Letters: Literacy, Power, and the Transmitters of Early Christian Literature*. Oxford: Oxford University Press: 55–61, 153–57.

Hamilton, N.T., N.M. Swerdlow, and G.J. Toomer. 1987. "The Canobic Inscription: Ptolemy's Earliest Work." *From Ancient Omens to Statistical Mechanics: Essays on the Exact Sciences Presented to Asger Aaboe*. J.L. Berggren and B.R. Goldstein, eds. Copenhagen: University Library: 55–73.

Hankinson, R.J. 1991. *Galen: On the Therapeutic Method*. Oxford: Clarendon Press.

————. 1992. "Galen's Philosophical Eclecticism." *Aufstieg und Niedergang der römischen Welt* 2.36.5: 3505–22.

————. 1994. "Galen's Concept of Scientific Progress." *Aufstieg und Niedergang der römischen Welt* 2.37.2: 1776–89.

————, ed. 2008. *The Cambridge Companion to Galen*. Cambridge: Cambridge University Press.

Hannam, James. 2009. *God's Philosophers: How the Medieval World Laid the Foundations of Modern Science*. London: Icon.

Hanson, Carl A. 1989. "Were There Libraries in Roman Spain?" *Libraries & Culture* 24.2 (Spring): 198–216.

Harich-Schwarzbauer, Henriette. 2011. *Hypatia: die spätantiken Quellen: eingeleitet, kommentiert und interpretiert*. Bern: Lang.

Harig, G. 1971. "Zum Problem « Krankenhaus » in Der Antike." *Klio* 53: 179–95.

Harris, William V. 1989. *Ancient Literacy*. Cambridge, Massachusetts: Harvard University Press.

Harrison, Peter. 1998. *The Bible, Protestantism, and the Rise of Natural Science*. Cambridge: Cambridge University Press.

———. 2001. "Curiosity: Forbidden Knowledge, and the Reformation of Natural Philosophy in Early Modern England." *Isis* 92.2 (June): 265–90.

Harrison, S.J. 2000. *Apuleius: A Latin Sophist*. Oxford: Oxford University Press.

Haskins, Charles Homer. 1923. *The Rise of Universities*. New York: H. Holt and Co.

Hastings, Rashdall. 1987. *The Universities of Europe in the Middle Ages*, rev. ed., 3 vols. (revised by F. M. Powicke and A. B. Emden). Oxford: Clarendon Press.

Hauge, Matthew Ryan, and Andrew Pitts, eds. 2016. *Ancient Education and Early Christianity*. New York: T & T Clark.

Healy, John. 1999. *Pliny the Elder on Science and Technology*. Oxford: Oxford University Press.

Hein, Wolfgang. 2012. *Mathematik im Altertum: von Algebra bis Zinseszins*. Darmstadt: Wissenschaftliche Buchgesellschaft.

Hemelrijk, Emily. 1999. *Matrona Docta: Educated Women in the Roman Élite from Cornelia to Julia Domna*. New York: Routledge.

Hezser, Catherine. 2001. *Jewish Literacy in Roman Palestine*. Tubingen: Mohr Siebeck.

Hin, Saskia. 2007. "Class and Society in the Cities of the Greek East: Education During the Ephebia." *Ancient Society* 37: 141–66.

Hornblower, Simon, and Antony Spawforth, eds. 2012. *The Oxford Classical Dictionary*. 4th ed. Oxford: Oxford University Press. (Assistant ed., Esther Eidinow.)

Horsfall, Nicholas. 1979. "Doctus Sermones Utriusque Linguae?" *Echos du Monde Classique* 23.3 (October): 79–95.

Houston, George W. 2002. "The Slave and Freedman Personnel of Public Libraries in Ancient Rome." *Transactions of the American Philological Association* 132.1–2 (Autumn): 139–176.

———. 2009. "Papyrological Evidence for Book Collections and Libraries in the Roman Empire." [in Johnson and Parker 2009: 233–67]

Huby, Pamela, and Gordon Neal, eds. 1989. *The Criterion of Truth: Essays Written in Honour of George Kerferd Together with a Text and Translation (with Annotations) of Ptolemy's* On the Kriterion *and* Hegemonikon. Liverpool: Liverpool University Press.

Hueber, Friedmund, and Volker Strocka. 1975. "Die Bibliothek des Celsus: Eine Prachtfassade in Ephesos und das Problem ihrer Wiederaufrichtung." *Antike Welt* 6.

Hulskamp, Maithe. 2012. "Space and the Body: Uses of Astronomy in Hippocratic Medicine." *Medicine and Space: Body, Surroundings and Borders in Antiquity and the Middle Ages*. Patricia Baker, Han Nijdam, and Karine van 't Land, eds.: 149–69.

Humphrey, John, ed. 1991. *Literacy in the Roman World*. Ann Arbor, Michigan: Journal of Roman Archaeology.

Hutchinson, D. S. 1988. "Doctrines of the Mean and the Debate Concerning Skills in Fourth-Century Medicine, Rhetoric and Ethics." *Apeiron: A Journal for Ancient Philosophy and Science* 21.2 (Summer): 17–52.

Inwood, Brad, ed. 2003. *The Cambridge Companion to the Stoics*. New York: Cambridge University Press.

Irby-Massie, Georgia. 1993. "Women in Ancient Science." *Women's Power, Man's Game: Essays on Classical Antiquity in Honor of Joy K. King*. Mary DeForest, ed. Wauconda, Illinois: Bochazy-Carducci Publishers: 354–72.

Irby-Massie, Georgia, and Paul Keyser. 2002. *Greek Science of the Hellenistic Era: A Sourcebook*. London: Routledge.

Iskandar, Albert. 1976. "An Attempted Reconstruction of the Late Alexandrian Medical Curriculum." *Medical History* 20: 235–58.

———. 1988. *Galen on Examinations by Which the Best Physicians Are Recognized*. Berlin: Akademie-Verlag.

Jackson, Ralph. 1988. *Doctors and Diseases in the Roman Empire*. Norman: University of Oklahoma Press.

———. 1993. "Roman Medicine: The Practitioners and Their Practices." *Aufstieg und Niedergang der römischen Welt* 2.37.1: 79–101.

Jacobs, Andrew. 2011. "'What Has Rome to Do with Bethlehem?': Cultural Capital(s) and Religious Imperialism in Late Ancient Christianity." *Classical Receptions Journal* 3.1: 29–45.

Jaeger, Mary. 2002. "Cicero and Archimedes' Tomb." *Journal of Roman Studies* 92: 49–61.

James, Peter, and Nick Thorpe. 1994. *Ancient Inventions*. New York: Ballantine.

Johnson, Lora Lee. 1984. *The Hellenistic and Roman Library: Studies Pertaining to Their Architectural Form*. Dissertation, Ph.D. (Brown University).

Johnson, William and Holt Parker, eds. 2009. *Ancient Literacies: The Culture of Reading in Greece and Rome* (New York: Oxford University Press).

Jones, Alexander. 1994. "The Place of Astronomy in Roman Egypt." [in T. Barnes 1994: 25–51]

———. 2006. "The Astronomical Inscription from Keskintos, Rhodes." *Mediterranean Archaeology and Archaeometry* 6.3: 215–22.

Jones, C.P. 1989. "Eastern Alimenta and an Inscription of Attaleia." *Journal of Hellenic Studies* 109: 189–91.

Jones, Christopher. 2009. "Books and Libraries in a Newly-Discovered Treatise of Galen." *Journal of Roman Archaeology* 22.1: 390–97.

Joost-Gaugier, Christiane. 2006. *Measuring Heaven: Pythagoras and His Influence on Thought and Art in Antiquity and the Middle Ages*. Ithaca, New York: Cornell University Press.

Joyal, Mark, Iain McDougall, and John Yardley. 2009. *Greek and Roman Education: A Sourcebook*. New York: Routledge.

Judge, E.A. 1983. "The Reaction against Classical Education in the New Testament." *Journal of Christian Education* 77.1: 7–14.

Kah, Daniel and Peter Scholz, eds. 2004. *Das hellenistische Gymnasion*. Berlin: Akademie Verlag.

Kalligas, Paul. 2004. "Platonism in Athens during the First Two Centuries A.D.: An Overview." *Rhizai: A Journal for Ancient Philosophy and Science* 2.2: 37–56.

Kaster, Robert. 1988. *Guardians of Language: The Grammarian and Society in Late Antiquity*. Berkeley: University of California Press.

Kenny, Neil. 1998. *Curiosity in Early Modern Europe*. Wiesbaden: Harrassowitz.

———. 2004. *The Uses of Curiosity in Early Modern France and Germany*. New York: Oxford University Press.

Keyser, Paul, and Georgia Irby-Massie. 2008. *The Encyclopedia of Ancient Natural Scientists*. New York: Routledge.

Kidd, I. G. 1988. *Posidonius II: The Commentary*. Cambridge: Cambridge University Press.

King, Helen. 1986. "Agnodike and the Profession of Medicine." *Proceedings of the Cambridge Philological Society* 212: 53–77.

Klauck, Hans-Josef. 2003. *The Religious Context of Early Christianity: A Guide to Graeco-Roman Religions*. Minneapolis: Fortress.

Kleijwegt, Marc. 1991. *Ancient Youth: The Ambiguity of Youth and the Absence of Adolescence in Greco-Roman Society* (Amsterdam: J. C. Gieben).

Knauber, Adolf. 1968. "Das Anliegen der Schule des Origenes zu Cäsarea." *Münchener Theologische Zeitschrift* 19.3: 182–203.

Knorr, Wilbur. 1989. *Textual Studies in Ancient and Medieval Geometry*. New York: Birkhauser.

———. 1993. "*Arithmêtikê stoicheiôsis*: On Diophantus and Hero of Alexandria." *Historia Mathematica* 20.2 (May): 180–92.

Koertge, Noretta, ed. 2008. *The New Dictionary of Scientific Biography*. Detroit : Charles Scribner's Sons.

Kollesch, Jutta. 1979. "Ärztliche Ausbildung in Der Antike." *Klio* 61.2: 507–13.

Kollesch, Jutta, and Diethard Nickel, eds. 1993. *Galen und das hellenistische Erbe: Verhandlungen des IV. internationalen Galen-Symposiums veranstaltet vom Institut für Geschichte der Medizin am Bereich Medizin (Charité) der Humboldt-Universität zu Berlin, 18.-20. September 1989.* Stuttgart: Franz Steiner.

König, Jason. 2005. *Athletics and Literature in the Roman Empire*. New York: Cambridge University Press.

———. 2009. "Education." *A Companion to Ancient History*. Andrew Erskine, ed. Malden, MA: Wiley-Blackwell: 392–402.

———. 2013. *Ancient Libraries*. Jason König, Katerina Oikonomopoulou, and Greg Woolf, eds. Cambridge: Cambridge University Press.

König, Jason, and Greg Woolf, eds. 2013. *Encyclopaedism from Antiquity to the Renaissance*. Cambridge: Cambridge University Press.

Korpela, Jukka. 1987. *Das Medizinalpersonal im antiken Rom: Eine sozialgeschichte Untersuchung*. Helsinki: Suomalainen Tiedeakatemia.

Kozak, Lynn. 2013. "Greek Government and Education: Re-Examining the *ephêbeia*." *A Companion to Ancient Greek Government*. Hans Beck, ed. Chichester: Wiley-Blackwell: 302–16.

Kudlien, Fridolf. 1970. "Medical Education in Classical Antiquity." *The History of Medical Education*, ed. by C.D. O'Malley. Berkeley: University of California Press: 3–37.

———. 1976. "Medicine as a 'Liberal Art' and the Question of the Physician's Income." *Journal of the History of Medicine and Allied Sciences* 31.4 (October): 448–459.

———. 1985. "Jüdische Ärzte im römischen Reich." *Medizinhistorisches Journal* 20.1/2: 36–57.

Kühn, Karl Gottlob. 1821–1833. *Claudii Galeni Opera Omnia*. [= *Medicorum Graecorum Opera Quae Exstant* 1–20]

Künzl, Ernst. 1995. "Ein archäologisches Problem: Gräber römischer Chirurginnen." [in van der Eijk, Horstmanshoff and Schrijvers 1995: 1.309–19]

Kyriakis, M. J. 1971. "The University: Origin and Early Phases in Constantinople." *Byzantion* 41: 161–182.

La Bua, Giuseppe. 2010. "Patronage and Education in Third-Century Gaul: Eumenius' Panegyric for the Restoration of the Schools." *Journal of Late Antiquity* 3.2: 300–15.

Laes, Christian. 2007. "School-teachers in the Roman Empire: A Survey of the Epigraphical Evidence." *Acta Classica* 50: 109–27.

Lamotte, Hélène. 2007. "L'œuvre de Trajan en faveur des enfants de la plèbe romaine: un essai de politique nataliste?" *Mélanges de lÉcole fraiņçaise de Rome: Antiquité* 119.1: 189–224.

Lapin, Hayim. 1996. "Jewish and Christian Academies in Roman Palestine: Some Preliminary Observations." *Caesarea Maritima: A Retrospective after Two Millennia*. Avner Raban and Kenneth Holum, eds. New York: E.J. Brill: 496–511.

Lawson, R.P. 1957. *Origen: The Song of Songs, Commentary and Homilies*. Westminster, Maryland: The Newman Press.

Lee, Desmond. 1973. "Science, Philosophy and Technology in the Greco-Roman World [Part I]" *Greece and Rome* (2nd ser.) 20.1 (April): 65–78.

Lee, Michelle. 2006. *Paul, the Stoics, and the Body of Christ*. New York: Cambridge University Press.

Lehoux, Daryn. 2007. *Astronomy, Weather, and Calendars in the Ancient World: Parapegmata and Related Texts in Classical and Near Eastern Societies*. New York: Cambridge University Press.

———. 2012. *What Did the Romans Know? An Inquiry into Science and Worldmaking*. Chicago: University of Chicago Press.

Levick, Barbara. 2002. "Women, Power, and Philosophy at Rome and Beyond." [in Clark and Rajak 2002: 133–55]

———. 2007. *Julia Domna, Syrian Empress*. New York: Routledge.

Levin, Flora. 2009. *Greek Reflections on the Nature of Music*. Cambridge: Cambridge University Press.

Lewis, Charlton, and Charles Short. 1879. *A Latin Dictionary*. Oxford: Clarendon Press.

Lewis, Naphtali. 1963. "The Non-Scholar Members of the Alexandrian Museum." *Mnemosyne* 16.3 (series 4): 257–61.

Lewis, Naphtali, and Meyer Reinhold, eds. 1990. *Roman Civilization: Selected Readings*, 3rd ed. 2 vols. New York: Columbia University Press.

"Library." 2005. *New Pauly – Brill's Encyclopedia of the Ancient World* 7: 498–511.

Liddell, Henry, and Robert Scott. 1996. *A Greek-English Lexicon*. 9th ed., with a revised supplement. Oxford: Clarendon Press.

Lloyd, G.E.R. 1973. *Greek Science After Aristotle*. New York: W. W. Norton.

———. 1983. *Science, Folklore, and Ideology: Studies in the Life Sciences in Ancient Greece*. New York: Cambridge University Press.

———. 2005. "Mathematics as a Model of Method in Galen." [in Sharples 2005: 110–30]

Long, A.A. 1982. "Astrology: Arguments Pro and Contra." *Science and Speculation: Studies in Hellenistic Theory and Practice*. Barnes, Jonathan, Jacques Brunschwig, Myles Burnyeat, and Malcolm Schofield, eds. Cambridge: Cambridge University Press: 165–92.

———. 1988. "Ptolemy *On the Criterion*: An Epistemology for the Practicing Scientist." [in Dillon & Long 1988: 176–207]

———. 2002. *Epictetus: A Stoic and Socratic Guide to Life*. New York: Oxford University Press.

Losev, A. 2012. "'Astronomy' or 'Astrology': A Brief History of an Apparent Confusion." *Journal of Astronomical History and Heritage* 15: 42–46.

Lougee, David Gilman. 1972. *The Concept of the Natural Scientist in Seventeenth-Century Defenses of Science in England*. Dissertation, Ph.D. (University of Michigan, Ann Arbor).

Maass, Ernest. 1958. *Commentariorum in Aratum Reliquiae*. Berlin: Weidmanns.

MacLeod, Roy, ed. 2000. *The Library of Alexandria: Centre of Learning in the Ancient World*. London: I.B. Taurus.

MacMullen, Ramsay. 1984. *Christianizing the Roman Empire (A.D. 100–400)*. New Haven: Yale University Press.

———.1997. *Christianity and Paganism in the Fourth to Eighth Centuries*. New Haven: Yale University Press.

———. 2009. *The Second Church: Popular Christianity A.D. 200–400*. Atlanta: Society of Biblical Literature.

Makowiecka, Elżbieta. 1978. *The Origin and Evolution of Architectural Form of Roman Library*. Warsaw. WUW.

Markopoulos, Athanasios. 2008. "Education." *The Oxford Handbook of Byzantine Studies*. Elizabeth Jeffreys, John Haldon, and Robin Cormack, eds. Oxford: Oxford University Press: 785–95.

Marrou, Hénri. 1956. *A History of Education in Antiquity*, 3rd ed. New York: Sheed and Ward. [tr. by George Lamb of *Histoire de l'éducation dans l'antiquité*, 3rd ed. Paris: Editions du Seuil: 1955]

———. 1964. *Histoire de l'éducation dans l'antiquité*, 6th ed. Paris: Éditions du Seuil.

———. 1981. "Education and Rhetoric." *The Legacy of Greece: A New Appraisal*. Moses Finley, ed. Oxford: Clarendon Press: 185–201.

Marshall, Anthony J. 1976. "Library Resources and Creative Writing at Rome." *Phoenix* 30.3 (Autumn): 252–64.

Mattern, Susan. 2008. *Galen and the Rhetoric of Healing*. Baltimore: Johns Hopkins University Press.

———. 2013. *The Prince of Medicine: Galen in the Roman World*. Oxford: Oxford University Press.

Maurice, Lisa. 2013. *The Teacher in Ancient Rome: The Magister and His World*. Lanham, Maryland: Lexington Books.

May, Margaret Tallmadge. 1968. *Galen: On the Usefulness of the Parts of the Body*. Ithaca, New York: Cornell University Press.

McGrayne, Sharon Bertsch. 2011. *The Theory That Would Not Die: How Bayes' Rule Cracked the Enigma Code, Hunted down Russian Submarines, and Emerged Triumphant from Two Centuries of Controversy*. New Haven: Yale University Press.

McKenzie, Judith. 2007. *The Architecture of Alexandria and Egypt, c. 300 B.C. to A.D. 700*. New Haven, Connecticut: Yale University Press.

McKirahan, Voula Tsouna. 1994. "The Socratic Origins of the Cynics and Cyrenaics." *The Socratic Movement*. Paul A. Vander Waerdt, ed. Ithaca:

Cornell University Press: 367–91.

Meacham, Tirzah. 2000. "Halakhic Limitations on the Use of Slaves in Physical Examinations." *From Athens to Jerusalem: Medicine in Hellenized Jewish Lore and in Early Christian Literature: Papers of the Symposium in Jerusalem, 9–11 September 1996*. Samuel Kottek and Manfred Horstmanshoff, eds. Rotterdam: Erasmus: 33–48.

Meggitt, Justin. 2010. "Popular Mythology in the Early Empire and the Multiplicity of Jesus Traditions." *Sources of the Jesus Tradition: Separating History from Myth*. R. Joseph Hoffmann, ed. Amherst, New York: Prometheus Books: 55–80.

Meunier, Louise. 1997. *Le Médicin Grec dans la Cité Hellénistique*. Dissertation, M.A. (Université Laval, Québec City, Canada).

Millard, Alan. 2000. *Reading and Writing in the Time of Jesus*. New York: New York University Press.

———. 2003a. "Literacy in the Time of Jesus: Could His Words Have Been Recorded in His Lifetime?" *Biblical Archaeology Review* 29.4 (July/August): 36–45.

———. 2003b. "Alan Millard Responds." *Biblical Archaeology Review* 29.6 (November/December): 14–16.

Morford, Mark. 2002. *The Roman Philosophers: From the Time of Cato the Censor to the Death of Marcus Aurelius*. New York: Routledge.

Morgan, Teresa. 1998. *Literate Education in the Hellenistic and Roman Worlds*. New York: Cambridge University Press.

———. 2007. "Rhetoric and Education." *A Companion to Greek Rhetoric*. Ian Worthington, ed. Oxford: Blackwell: 303–319.

———. 2011. "Ethos: The Socialization of Children in Education and Beyond." *A Companion to Families in the Greek and Roman Worlds*. Beryl Rawson, ed. Oxford: Blackwell: 504–520.

Morison, Ben. 2008. "Logic." *The Cambridge Companion to Galen*, ed. R.J. Hankinson. New York: Cambridge University Press.

Mudry, P. 1986. "Science et conscience: Réflexions sur le discours scientifique à Rome." *Études de Lettres* 1: 75–86.

Mullen, Alex, and Patrick James. 2012. *Multilingualism in the Graeco-Roman Worlds*. New York: Cambridge University.

Nahin, Paul. 1998. *An Imaginary Tale: The Story of √-1*. Princeton, New Jersey: Princeton University Press.

Nelson, C.A. 1983. *Financial and Administrative Documents from Roman Egypt*. Berlin: Staatliche Museen Preussischer Kulturbesitz, 1983.

Nesselrath, Heinz-Günther. 2012. "Did It Burn or Not? Caesar and the Great Library of Alexandria: A New Look at the Sources." *Quattuor Lustra: Papers Celebrating the 20th Anniversary of the Re-establishment of Classical Studies at the University of Tartu*. Ivo Volt and Janika Päll, eds. Tartu: Tartu Ülikooli Kirjastus: 56–74.

Netz, Reviel. 2002. "Greek Mathematicians: A Group Picture." [in Tuplin & Rihll 2002: 196–216]

———. 2003. "The Goal of Archimedes' *Sand Reckoner*." *Apeiron: A Journal for Ancient Philosophy and Science* 36.4 (December): 251–90.

———. 2011. "The Bibliosphere of Ancient Science (Outside of Alexandria): A Preliminary Survey." *NTM: International Journal of History & Ethics of Natural Sciences Technology & Medicine* 19.3: 239–69.

Netz, Reviel, and William Noel. 2007. *The Archimedes Codex: How a Medieval Prayer Book is Revealing the True Genius of Antiquity's Greatest Scientist*. Philadelphia: Da Capo Press.

Netz, Reviel, and Fabio Acerbi and Nigel Wilson. 2004. "Towards a Reconstruction of Archimedes' Stomachion," Sciamus: Sources and Commentaries in Exact Sciences 5, 2004: 67–99.

Newmyer, Stephen Thomas. 1996. "Talmudic Medicine and Greco-Roman Science: Cross-Currents and Resistance." *Aufstieg und Niedergang der römischen Welt* 2.37.3: 2895–2911.

Nicholls, Matthew. 2010. "Parchment Codices in a New Text of Galen." *Greece & Rome* 2010 (Ser. 2) 57.2: 378–86.

———. 2011. "Galen and Libraries in the *Peri Alypias*." *Journal of Roman Studies* 101: 123–42.

Nickel, Diethard. 1979. "Berufsvorstellungen über weibliche Medizinalpersonen in der Antike." Klio 61.2: 515–18.

Nilsson, Martin. 1955. *Die Hellenistische Schule*. Münich: Verlag C.H. Beck.

Noble, Joseph, and Derek de Solla Price. 1968. "The Water Clock in the Tower of the Winds." *American Journal of Archaeology* 72.4 (October): 345–55.

Nutton, Vivian. 1971a. "L. Gellius Maximus, Physician and Procurator." *Classical Quarterly* 21.1: 262–72.

———. 1971b. "Two Notes on Immunities: *Digest* 27, 1, 5, 10 and 11." *Journal of Roman Studies* 61: 52–63.

———. 1973. "The Chronology of Galen's Early Career." *The Classical Quarterly* 23.1 (May): 158–71.

———. 1975. "Museums and Medical Schools in Classical Antiquity." *History of Education* 4.1 (Spring): 3–15.

———. 1977. "*Archiatri* and the Medical Profession in Antiquity." *Papers of the British School at Rome* 45: 191–226.

———. 1981. "Continuity or Rediscovery? The City Physician in Classical Antiquity and Mediaeval Italy." *The Town and State Physician in Europe from the Middle Ages to the Enlightenment*. Andrew Russell, ed. Wolfenbüttel: Herzog August Bibliothek: 9–46.

———. 1985. "Murders and Miracles: Lay Attitudes towards Medicine in Classical Antiquity." *Patients and Practitioners: Lay Perceptions of Medicine in Pre-Industrial Society*. Roy Porter, ed. New York: Cambridge University Press.

———. 1993. "Galen and Egypt." [in Kollesch & Nickel 1993: 11–31]

———.1995. "The Medical Meeting Place." [in van der Eijk, Horstmanshoff and Schrijvers 1995: 1.3–25]

———. 1997. "Galen on Theriac: Problems of Authenticity." *Galen on Pharmacology: Philosophy, History, and Medicine: Proceedings of the Vth International Galen Colloquium, Lille, 16–18 March 1995*. Armelle Debru, ed. Leiden: Brill: 133–51.

———. 1999. *Galen: On My Own Opinions*. Berlin: Akademie Verlag. [= *Corpus Medicorum Graecorum* 5.3.2]

———. 2004. *Ancient Medicine*. London: Routledge.

O'Keefe, Tim. 2010. *Epicureanism*. Berkeley: University of California Press.

Oleson, John Peter. 2004. "*Well-Pumps for Dummies*: Was There a Roman Tradition of Popular Sub-Literary Engineering Manuals?" *Problemi di macchinismo in ambito Romano macchine idrauliche nella letteratura tecnica, nelle fonti storiografiche e nelle evidenze archeologiche di età imperiale*. Franco Minonzio, ed. Coma, Italy: Comune di Como, Assessorato alla Cultura, Musei Civici: 65–86.

———, ed. 2008. *The Oxford Handbook of Engineering and Technology in the Classical World*. Oxford: Oxford University Press.

Oliver, James. 1970. *Marcus Aurelius: Aspects of Civic and Cultural Policy in the East*. Hesperia Supplements 13. Princeton: American School of Classical Studies at Athens.

———. 1977. "The Diadoche at Athens under the Humanistic Emperors." *The American Journal of Philology* 98.2 (Summer): 160–78.

———. 1981. "Marcus Aurelius and the Philosophical Schools at Athens." *The American Journal of Philology* 102.2 (Summer): 213–25.

Osborn, Eric. 2005. *Clement of Alexandria*. New York: Cambridge University Press.

Ostler, Nicholas. 2007. *Ad Infinitum: A Biography of Latin*. New York: Walker & Co.

Pack, Edgar. 1989. "Sozialgeschichtliche Aspekte des Fehlens einer 'Christlichen' Schule in der römischen Kaiserzeit." *Religion und Gesellschaft in der römischen Kaiserzeit: Kolloquium zu Ehren von Friedrich Vittinghoff*. Köln: Böhlau: 185–263.

Page, Marie-Michelle. 2009. "La politique socio-agraire de l'empereur Nerva (96–98)." *Mélanges de lÉcole fraiņçaise de Rome: Antiquité* 121.1: 209–40.

Pailler, Jean-Marie, and Pascal Payen, eds. 2004. *Que reste-t-il de l'éducation classique? Relire «le Marrou» Histoire de l'éducation dans l'Antiquité*. Toulouse: Presses Universitaires du Mirail.

Penella, Robert. 2011–2012. "The *progymnasmata* in Imperial Greek Education." *The Classical World* 105.1: 77–90.

Parker, H.M. 1997. "Women Doctors in Greece, Rome, and the Byzantine Empire." *Women Healers and Physicians: Climbing a Long Hill.* L.R. Furst, ed. Lexington, Kentucky: University Press of Kentucky: 131–50.

Parks, E. Patrick. 1945. *The Roman Rhetorical Schools as a Preparation for the Courts under the Early Empire* (published as *The Johns Hopkins University Studies in Historical and Political Science* 63.2). Baltimore: The Johns Hopkins Press.

Parsons, Edward. 1952. *The Alexandrian Library, Glory of the Hellenic World: Its Rise, Antiquities, and Destructions.* Amsterdam: Elsevier Press.

Parsons, P.J. 1976. "Petitions and a Letter: The Grammarian's Complaint." *Collectanea Papyrologica: Texts Published in Honor of H.C. Youtie* 2: 409–46.

Patterson, John. 1987. "Crisis: What Crisis? Rural Change and Urban Development in Imperial Appennine Italy." *Papers of the British School at Rome* 55: 115–46.

Pearcy, Lee. 1993. "Medicine and Rhetoric in the Period of the Second Sophistic." *Aufstieg und Niedergang der römischen Welt* 2.37.1: 445–56.

Pedersen, Olaf. 1997. *The First Universities: Studium Generale and the Origins of University Education in Europe.* Cambridge: Cambridge University Press.

Perrin-Saminadayar, Eric. 2004. "À chacun son dû: la rémunération des maîtres dans le monde grec classique et hellénistique." [in Pailler & Payen 2004: 307–18]

Petermandl, Werner. 2014. "Growing up with Greek Sport: Education and Athletics." *A Companion to Sport and Spectacle.* Paul Christesen and Donald Kyle, eds. Chichester: Wiley-Blackwell: 236–245.

Peters, Siegwart. 2010. "Römische Valetudinaria an der Rheinfront: Die stationäre Betreuung des kranken Legionärs zur frühen Kaiserzeit." *Würzburger Medizinhistorische Mitteilungen* 29: 158–93.

Pitts, Lynn, and J.K. St. Joseph, eds. 1985. *Inchtuthil: The Roman Legionary Fortress Excavations, 1952–65.* London: Society for the Promotion of Roman Studies.

Plant, I.M. 2004. *Women Writers of Ancient Greece and Rome: An Anthology*. Norman: University of Oklahoma Press.

Platt, David Stuart. 2008. *A Cultural Studies Approach to Roman Public Libraries: Social Negotiation, Changing Spaces, and Euergetism*. Dissertation, Ph.D. (Stanford University).

Pomeroy, Sarah. 1977. "*Technikai kai Mousikai*: The Education of Women in the Fourth Century and in the Hellenistic Period." *American Journal of Ancient History* 2.1: 51–68.

———. 1995. *Goddesses, Whores, Wives, and Slaves: Women in Classical Antiquity*. New York: Schocken Books.

———. 2013. *Pythagorean Women: Their History and Writings*. Baltimore: Johns Hopkins University Press.

Poulakos, Takis. 1997. "Educational Program." *Speaking for the Polis: Isocrates' Rhetorical Education*. Columbia, South Carolina: University of South Carolina Press: 93–115.

Press, Ludwika. 1988. "Valetudinarium at Novae and Other Roman Danubian Hospitals." *Archeologia* 39: 69–89.

Prince, Susan. 2006. "Socrates, Antisthenes, and the Cynics." *A Companion to Socrates*. Sara Ahbel-Rappe and Rachana Kamtekar, eds. Oxford: Blackwell: 75–92.

Quick, Laura. 2014. "Recent Research on Ancient Israelite Education: A Bibliographic Essay." *Currents in Biblical Research* 13.1 (October): 9–33.

Ramsay, Hazel. 1936. "Government Relief during the Roman Empire." *The Classical Journal* 31.8 (May): 479–88.

Rasimus, Tuomas, Troels Engberg-Pedersen, and Ismo Dunderberg, eds. 2010. *Stoicism in Early Christianity*. Grand Rapids: Baker Academic.

Rawson, Elizabeth. 1985. *Intellectual Life in the Late Roman Republic*. Baltimore, Maryland: Johns Hopkins University Press.

Remes, Pauliina. 2008. *Neoplatonism*. Berkeley: University of California Press.

Remus, Harold. 1983. *Pagan-Christian Conflict over Miracle in the Second Century*. Cambridge, Massachusetts: Philadelphia Patristic Foundation.

Reese, William J. 2005. *America's Public Schools: From the Common School to "No Child Left Behind."* Baltimore, Md.: Johns Hopkins University Press.

Richards, E. Randolph. 2004. *Paul and First-Century Letter Writing: Secretaries, Composition, and Collection*. Downers Grove, Illinois: InterVarsity Press.

Richardson, L. 1977. "The Libraries of Pompeii." *Archaeology* 30.6 (November): 394–402.

Rihll, Tracey. 1999. *Greek Science*. Oxford: Oxford University Press.

———. 2002. "Introduction: Greek Science in Context." [in Tuplin & Rihll 2002: 1–21]

Robinson, Rodney. 1921. "The Roman School Teacher and His Reward." *Classical Weekly* 15.8 (December 5): 57–61.

Rocca, Julius. 2003. *Galen on the Brain: Anatomical Knowledge and Physiological Speculation in the Second Century A.D.* Leiden: Brill.

Rollston, Chris. 2010. *Writing and Literacy in the World of Ancient Israel: Epigraphic Evidence from the Iron Age*. Atlanta: Society of Biblical Literature.

Rosner, Fred. 1994. "Physicians in the Talmud." *Medicine in the Bible and the Talmud: Selections from Classical Jewish Sources*. Hoboken, New Jersey: Yeshiva University Press: 211–15.

Rossi, Paolo. 2001. *The Birth of Modern Science*. Oxford: Blackwell.

Rosumek, Peter. 1982. *Technischer Fortschritt und Rationalisierung im antiken Bergbau*. Bonn: Dr. Rudolf Habelt G.M.B.H.

Rowland, Ingrid, and Thomas Howe. 1999. *Vitruvius: Ten Books on Architecture*. Cambridge: Cambridge University Press.

Rubenstein, Jeffrey. 2003. *The Culture of the Babylonian Talmud*. Baltimore: The Johns Hopkins University Press.

Russell, D.A. 1989. "Arts and Sciences in Ancient Education." *Greece & Rome* 36.2 (October): 210–25.

Russo, Lucio. 2003. *The Forgotten Revolution: How Science Was Born in 300 B.C. and Why It Had to Be Reborn*, 2nd ed. Berlin: Springer.

Safrai, Shmuel. 1969. "Elementary Education in the Talmudic Period." *Jewish Society through the Ages*. H.H. Ben-Sasson and S. Ettinger, eds. New York: Schocken Books: 148–69.

Sandnes, Karl Olav. 2009. *The Challenge of Homer: School, Pagan Poets and Early Christianity*. London: T & T Clark.

Sarton, George. 1959. *A History of Science II: Hellenistic Science and Culture in the Last Three Centuries B.C.* Cambridge, Massachusetts: Harvard University Press.

Scarborough, John. 1968. "Roman Medicine and the Legions: A Reconsideration." *Medical History* 12: 254–61.

———. 1970. "Romans and Physicians." *The Classical Journal* 65.7: 296–306.

Schmidt, Thomas, and Pascale Fleury, eds. 2011. *Perceptions of the Second Sophistic and Its Times*. Toronto: University of Toronto Press.

Schürmann, Astrid. 1991. *Griechische Mechanik und antike Gesellschaft: Studien zur staatlichen Förderung einer technischen Wissenschaft*. Stuttgart: Franz Steiner.

Seddon, Keith. 2005. *Epictetus' Handbook and the Tablet of Cebes: Guides to Stoic Living*. New York: Routledge.

Selinger, Reinhard. 1999. "Experimente mit dem Skalpell am menschlichen Körper in der griechisch-römischen Antike." *Saeculum* 50.1: 29–47.

Sellars, John. 2006. *Stoicism*. Berkeley: University of California Press.

Sergueenkova, Valeria. 2009. *Natural History in Herodotus'* Histories. Dissertation, Ph.D. (Harvard University).

Sherwin-White, A.N. 1973. *The Roman Citizenship*, 2nd ed. Oxford: Clarendon Press.

Simms, D.L. 1990. "The Trail for Archimedes's Tomb." *Journal of the Warburg and Courtauld Institutes* 53: 281–86.

Singer, Charles Joseph. 1956. *Galen On Anatomical Procedures*. London: Oxford University Press.

Skeat, T.C. 1982. "The Length of the Standard Papyrus Roll and the Cost-Advantage of the Codex." *Zeitschrift für Papyrologie und Epigraphik* 45: 169–75.

Smith, A.M. 1999. *Ptolemy and the Foundations of Ancient Mathematical Optics: A Source Based Guided Study*. Philadelphia: American Philosophical Society.

Smith, Martin Ferguson. 1996. "The Philosophical Inscription of Diogenes of Oinoanda." *Denkschriften (Österreichische Akademie der Wissenschaften: Philosophisch-Historische Klasse)* Bd. 251. Wien: Verlag der Osterreichische Akademie der Wissenschaften.

Snyder, Jane McIntosh. 1989. *The Woman and the Lyre: Women Writers in Classical Greece and Rome*. Carbondale: Southern Illinois University Press.

Staden, Heinrich von. 1975. "Experiment and Experience in Hellenistic Medicine." *Bulletin of the Institute of Classical Studies of the University of London* 22: 178–99.

———. 1989. *Herophilus: The Art of Medicine in Early Alexandria*. Cambridge: Cambridge University Press.

———. 1995. "Anatomy as Rhetoric: Galen on Dissection and Persuasion." *Journal of the History of Medicine and Allied Sciences* 50.1 (January): 47–66.

———. 1997. "Galen and the 'Second Sophistic.'" *Aristotle and After*. Richard Sorabji, ed. London: University of London: 33–54.

Stahl, William Harris. 1962. *Roman Science: Origins, Development, and Influence to the Later Middle Ages*. Madison, Wisconsin: Greenwood.

———. 1971. *The Quadrivium of Martianus Capella: Latin Traditions in the Mathematical Sciences, 50 B.C.-A.D. 1250*. New York: Columbia University Press.

Staikos, Konstantinos. 2000. *The Great Libraries: From Antiquity to the Renaissance (3000 B.C. to A.D. 1600)*. New Castle, Delaware: Oak Knoll Press.

———. 2004. *The History of the Library in Western Civilization*, vols. 1 (Classical and Hellenistic period) and 2 (Roman period). Athens, Greece: Kotinos Publications.

Starr, Raymond. 1990. "The Used-Book Trade in the Roman World." *Phoenix* 44.2 (Summer): 148–57.

Strocka, Volker Michael. 1981. "Römische Bibliotheken." *Gymnasium* 88.3/4: 298–329.

Strohmaier, Gotthard. 1993. "Hellenistische Wissenschaft im neugefundenen Galenkommentar zur hippokratischen Schrift 'Über die Umwelt.'" [in Kollesch & Nickel 1993: 157–64]

Stückelberger, Alfred. 1965. *Senecas 88. Brief: Über Wert und Unwert der freien Künste*. Heidelberg: Carl Winter Universitätsverlag.

————. 1988. *Einführung in die antiken Naturwissenschaften*. Darmstadt: Wissenschaftliche Buchgesellschaft.

Tacoma, Laurens. 2008. "Urbanisation and Access to Land in Roman Egypt." *Feeding the Ancient Greek City*. R. Alston and O. van Nijf, eds. Leuven: Peeters: 85–108.

Taub, Liba. 2003. *Ancient Meteorology*. London: Routledge.

————. 2008. *Aetna and the Moon: Explaining Nature in Ancient Greece and Rome*. Corvallis, Oregon: Oregon State University Press.

————. 2010. "Translating the *Phainomena* across Genre, Language and Culture." *Writings of Early Scholars in the Ancient Near East, Egypt, Rome, and Greece: Translating Ancient Scientific Texts*. Annette Imhausen and Tanja Pommerening, eds. New York: de Gruyter: 119–37.

Taylor, Joan. 2003. *Jewish Women Philosophers of First-Century Alexandria: Philo's "Therapeutae" Reconsidered*. Oxford: Oxford University Press.

Temkin, Owsei. 1934. "Galen's 'Advice for an Epileptic Boy.'" *Bulletin of the History of Medicine* 2: 179–89.

Thompson, Dorothy. "Education and Culture in Hellenistic Egypt and beyond." 2007. *Escuela y literatura en Grecia antigua: Actas del simposio internacional, Universidad de Salamanca, 17–19 noviembre de 2004*. J. A. Fernández Delgado, Francisca Pordomingo Pardo, and Antonio Stramaglia, eds. Frosinone, Italy: Edizioni dell'Università degli studi di Cassino: 121–40.

Thompson, James Westfall. 1962. *Ancient Libraries*. London: Archon Books.

Thorsrud, Harald. 2009. Ancient Scepticism. Berkeley: University of California Press.

Tod, Marcus N. 1957. "Sidelights on Greek Philosophers." *The Journal of Hellenic Studies* 77.1: 132–141.

Todd, Robert. 1984. "Philosophy and Medicine in John Philoponus' Commentary on Aristotle's *De Anima*." *Dumbarton Oaks Papers* 38: 103–20.

Toner, Jerry. 2002. *Rethinking Roman History*. Cambridge: Oleander Press.

———. 2009. *Popular Culture in Ancient Rome*. Cambridge: Polity.

Too, Yun Lee, ed. 2001. *Education in Greek and Roman Antiquity*. Boston: Brill.

———. 2010. *The Idea of the Library in the Ancient World*. Oxford: Oxford University Press.

Tountas, Yannis. 2009. "The Historical Origins of the Basic Concepts of Health Promotion and Education: The Role of Ancient Greek Philosophy and Medicine." *Health Promotion International* 24.2 (June): 185–92.

Trapp, Michael. 1997. "On the Tablet of Cebes." *Aristotle and After*. Richard Sorabji, ed. London: University of London: 159–80.

———. 2007. *Philosophy in the Roman Empire: Ethics, Politics and Society*. Aldershot, England: Ashgate.

Trigg, Joseph Wilson. 1998. *Origen*. London: Routledge.

Tucci, Pier Luigi. 2008. "Galen's storeroom, Rome's Libraries, and the Fire of A.D. 192." *Journal of Roman Archaeology* 21.1: 133–49.

Tuplin, C.J., and T.E. Rihll, eds. 2002. *Science and Mathematics in Ancient Greek Culture*. Oxford: Oxford University Press.

Turner, E.G. 1980. *Greek Papyri: An Introduction*, 2nd ed. (Oxford: Clarendon Press).

Tybjerg, Karin. 2005. "Hero of Alexandria's Mechanical Treatises: Between Theory and Practice." [in Schürmann 2005: 204–26]

Ulrich, Jörg. 2012. "What Do We Know about Justin's 'School' in Rome?" *Zeitschrift für antikes Christentum* 16.1: 62–74.

Van Brummelen, Glen. 2009. *The Mathematics of the Heavens and the Earth: The Early History of Trigonometry*. Princeton, NJ: Princeton University Press.

———. 2013. *Heavenly Mathematics: The Forgotten Art of Spherical Trigonometry*. Princeton: Princeton University Press.

Vendries, Christophe. 2004. "La place de la musique dans l'éducation romaine selon Marrou: la vision d'un musicologue averti." [in Pailler & Payen 2004: 257–64]

Waithe, Mary Ellen. 1987. *Ancient Women Philosophers, 600 B.C.-500 A.D.* Hingham, Massachussetts: Kluwer Academic Publishers.

Walker, Jeffrey. 2011. *The Genuine Teachers of This Art: Rhetorical Education in Antiquity*. Columbia, SC: University of South Carolina Press.

Wallace, Sherman LeRoy. 1938. *Taxation in Egypt from Augustus to Diocletian*. Princeton, Princeton University Press.

Wallace-Hadrill, Andrew. 1983. *Suetonius: The Scholar and His Caesars*. London: Duckworth.

Walzer, Richard. 1949. *Galen on Jews and Christians*. London, Oxford University Press.

Walzer, Richard, and Michael Frede. 1985. *Galen: Three Treatises on the Nature of Science*. Indianapolis, Indiana: Hackett.

Wareh, Tarik. 2012. *The Theory and Practice of life: Isocrates and the Philosophers*. Cambridge, MA: Harvard University Press.

Warren, James. 2009. *The Cambridge Companion to Epicureanism*. New York: Cambridge University Press.

Watts, Edward J. 2006. *City and School in Late Antique Athens and Alexandria*. Berkeley: University of California Press.

White, Peter. 2009. "Bookshops in the Literary Culture of Rome." [in Johnson and Parker 2009: 268–87]

Whitehead, David, and P.H. Blyth. 2004. *Athenaeus Mechanicus:* On Machines *(Peri Mêchanêmatôn)*. Stuttgart: Franz Steiner Verlag.

Whitehorne, J.E.G. 1982. "The Ephebate and the Gymnasial Class in Roman Egypt." *The Bulletin of the American Society of Papyrologists* 19:3–4: 171–84.

Whitmarsh, Tim. 2001. *Greek Literature and the Roman Empire: The Politics of Imitation*. New York: Oxford University Press.

———. 2005. *The Second Sophistic*. Oxford: Oxford University Press.

———. 2013. *Beyond the Second Sophistic: Adventures in Greek Postclassicism*. Berkeley: University of California Press.

Wilmanns, Juliane. 1995. "Der Arzt in der römischen Armee der frühen und hohen Kaiserzeit." [in van der Eijk, Horstmanshoff and Schrijvers 1995: 1.171–88]

Winsbury, Rex. 2009. *The Roman Book: Books, Publishing and Performance in Classical Rome*. London: Duckworth.

Witty, Francis J. 1974. "Reference Books of Antiquity." *The Journal of Library History* 9.2 (April): 101–19.

Wolff, Catherine. 2015. *L'éducation dans le monde romain: du début de la République à la mort de Commode*. Paris: Picard.

Woodside, M. St. A. 1942. "Vespasian's Patronage of Education and the Arts." *Transactions of the American Philological Society* 73: 123–29.

Woolf, Greg. 1990. "Food, Poverty, and Patronage: The Significance of the Epigraphy of the Roman Alimentary Schemes in Early Imperial Italy." *Papers of the British School at Rome* 58: 197–228.

———. 2000. "Literacy." *The Cambridge Ancient History, Volume 11: The High Empire, A.D. 70–192*, 2nd ed. Alan Bowman, Peter Garnsey, and Dominic Rathbone, eds. Cambridge University Press: 875–97.

Wouters, Alfons. 2007. "Between the Grammarian and the Rhetorician: the κλίσις χρείας." *Bezugsfelder: Festschrift für Gerhard Petersmann zum 65. Geburtstag*. Veronika Coroleu Oberparleiter, Ingrid Hohenwallner, and Ruth Elisabeth Kritzer, eds. (Horn: Berger): 137–54.

Zilsel, Edgar. 1945. "The Genesis of the Concept of Scientific Progress." *Journal of the History of Ideas* 6.3: 325–49.

Index

ABOUT THE AUTHOR

Richard Carrier, PhD, is a philosopher and historian of antiquity, specializing in contemporary philosophy of naturalism and Greco-Roman philosophy, science, and religion, including the origins of Christianity. He is the author of numerous books, including *Sense and Goodness without God*, *Not the Impossible Faith*, *Hitler Homer Bible Christ*, *Proving History*, and *On the Historicity of Jesus*. They are all also available as audiobooks, read by Dr. Carrier. For more about Dr. Carrier and his work see www.richardcarrier.info.